ST(P) Technology Today
A Series for Technicians

# Physics for TEC Level II

## T L Lowe
BSc PhD
Lecturer, Department of Physical Sciences and Mathematics,
South Glamorgan Institute of Higher Education, Cardiff

## K Roe
MSc
Lecturer, Department of Physics,
Gloucestershire College of Arts and Technology

## G D Redford
CEng MIMechE MIProdE AMIED
Senior Lecturer, Mechanical and Production
Engineering Department, Wigan College of Technology

## D T Rees
BSc CEng MIEE
Dean of Faculty of Science & Technology,
Gwent College of Higher Education, Newport, Gwent

## A Greer
CEng MRAeS
Formerly Senior Lecturer,
Gloucestershire College of Arts and Technology

Stanley Thornes (Publishers) Ltd.

First published in 1983 by:
Stanley Thornes (Publishers) Ltd
Educa House
Old Station Drive
Leckhampton Road
CHELTENHAM GL53 0DN
England

British Library Cataloguing in Publication Data

Physics for TEC level II
1. Physics
I. Lowe, T.
530'.0246        OC21.2

ISBN 0–85950–315–1

Typeset by Tech-Set, Unit 3, Brewery Lane, Felling, Tyne & Wear
Printed and bound in Great Britain at The Pitman Press, Bath

ST(P) Technology Today
A Series for Technicians

# Physics
# for TEC
# Level II

Stanley Thornes (Publishers) will gladly supply a complete list of their Technology Today titles upon request.

# CONTENTS

# PREFACE

Physics Level II is a unit which is incorporated in many Technician Education Council (TEC) Certificate and Diploma programmes. This is because the unit contains a number of important physics concepts useful both to technicians continuing their studies in physics and to those studying physics as an ancillary subject.

This book covers the content of the standard unit TEC U81/846. Extra material has been added, in appropriate sections, in order that most of the original Physics II unit (U76/003) is covered.

Our aim has been to produce a book which will enable even the weakest student to acquire sufficient information and understanding of the content that he/she can pass the appropriate tests. To facilitate this a number of worked examples and exercises have been incorporated.

The book should enable the teacher to minimise the time spent on note-taking and so make available more time for practical work and class discussion. Also we aim to give a sufficiently detailed treatment that the absent student can easily catch up on any work that has been missed.

We wish to thank J.C. Siddons for his advice and constructive criticism and for the correction of a number of errors. Any errors that remain are our responsibility alone.

T L Lowe
K Roe
G D Redford
D T Rees
A Greer
1983

# 1

# LINEAR MOTION

## BASIC DEFINITION

### DISTANCE

Distance is the length measured along any path, straight or curved, between two points. The SI unit of distance is the metre (m).

### SPEED

Average speed is the distance moved in unit time. Hence

$$\bar{v} = \frac{s}{t} \qquad [1.1]$$

where     $s$ = distance moved (metres)

$t$ = time taken (seconds)

and     $\bar{v}$ = average speed (metres per second)

Neither speed nor distance involve any particular direction, and so they are *scalar quantities*.

### DISPLACEMENT

This is the direct distance between two points and it is specified by giving this distance and the direction in which it is measured. That is, both magnitude and direction are involved and hence displacement is a *vector quantity*. For example, point B is displaced 10 metres horizontally to the right of a point A.

### VELOCITY

The rate at which a body is displaced is called velocity so that

$$\bar{v} = \frac{s}{t} \qquad [1.2]$$

1

Equations [1.1] and [1.2] are identical, but [1.2] is concerned with vector quantities, i.e. velocity is a vector quantity.

## ACCELERATION

Acceleration is the rate of change of velocity. Hence

$$a = \frac{v - u}{t}$$    [1.3]

where   $a$ = linear acceleration (metres per second per second)

$v$ = final velocity (metres per second)

$u$ = initial velocity (metres per second)

$t$ = time taken (seconds)

### WORKED EXAMPLE 1

A car moving with a uniform acceleration of $2\,m\,s^{-2}$ has an initial velocity of $3\,m\,s^{-1}$ when first observed. Calculate: (a) its velocity 4 seconds later and  (b) the distance it travels during those 4 seconds.

*Solution*

We are given that $u = 3\,m\,s^{-1}$, $a = 2\,m\,s^{-2}$ and $t = 4\,s$

(a)   From equation [1.3]

$$a = \frac{v - u}{t}$$

Rearranging gives

$$v = u + at = 3 + 2 \times 4 = 11$$

Hence the velocity after 4 seconds is $11\,m\,s^{-1}$.

(b)   Rearranging equation [1.2] gives

$$s = \text{Average velocity} \times \text{Time}$$

$$= \tfrac{1}{2}(u + v) \times t$$

$$= \tfrac{1}{2}(3 + 11) \times 4 = 28$$

The distance travelled is 28 m.

*WORKED EXAMPLE 2*

A car at a certain instant has a speed of $15\,\mathrm{m\,s^{-1}}$. It is given a uniform retardation of $0.5\,\mathrm{m\,s^{-2}}$ until its speed is reduced to $10\,\mathrm{m\,s^{-1}}$. Determine the distance travelled during the retardation.

*Solution*

Retardation is a negative acceleration. We are given that $u = 15\,\mathrm{m\,s^{-1}}$, $a = -0.5\,\mathrm{m\,s^{-2}}$ and $v = 10\,\mathrm{m\,s^{-1}}$. Rearranging equation [1.3] gives

$$t = \frac{v-u}{a} = \frac{10-15}{-0.5} = \frac{-5}{-0.5} = 10\ \text{seconds}$$

Now

$$s = \tfrac{1}{2}(u+v) \times t = \tfrac{1}{2}(15+10) \times 10$$
$$= 125$$

The distance travelled is 125 m.

## RESULTANT AND COMPONENTS OF VECTORS _____

A *vector quantity* is one which has both magnitude and direction. Force and velocity are vector quantities.

The *resultant* of two vector quantities is the single vector which has the same effect as the combined vectors. Fig. 1.1 shows two vectors of magnitude 5 units and 3 units acting in different directions. The resultant vector is of magnitude 7.1 units in the direction shown. We add vector quantities together using the parallelogram rule:

> If two vectors (OA and OB) acting at a point (O) are represented in magnitude and direction by the adjacent sides of a parallelogram, then the resultant (OC) is represented in magnitude and direction by the diagonal of the parallelogram passing through the point.

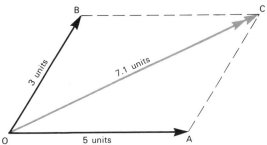

Fig. 1.1   Vector addition

It follows that a single vector quantity can be considered as *equivalent to two vectors* which we call the *components* of the original vector. We say a single vector can be *resolved* into its *component* vectors. In practice, we usually resolve into component directions at right angles.

Fig. 1.2 shows a vector $T$ acting at angle $\theta$ to the horizontal. Using trigonometry gives

Horizontal component     $H = T \cos \theta$                    [1.4]

Vertical component       $V = T \sin \theta$                    [1.5]

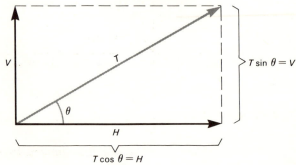

Fig. 1.2   Components of a vector

Note that vertical component $V$ has no effect in the direction of horizontal component $H$ (for example a vertical force has no effect in the horizontal direction) and vice versa. This is why we resolve into perpendicular directions.

*WORKED EXAMPLE 3*

A ship moves with velocity $5\,\text{m s}^{-1}\,\text{N}30°\text{E}$. Find the time taken to sail (a) 3 km East and (b) 3 km North.

*Solution*

Fig. 1.3 shows the vector diagram. We must find the northerly (N) and easterly (E) components of velocity.

(a)   From equation [1.4]

$$H = T \cos \theta$$

where          $H = E, \quad T = 5\,\text{m s}^{-1}, \quad \theta = 60°.$ So

$$E = 5 \cos 60 = 2.5$$

The ship travels easterly at $2.5\,\text{m s}^{-1}$. The time $t$ to travel 3 km East is given by

$$t = \frac{\text{Displacement}}{\text{Velocity}} = \frac{3000}{2.5} = 1200$$

The ship takes 1200 seconds to travel 3 km East of its original position. During this time it will also have travelled North.

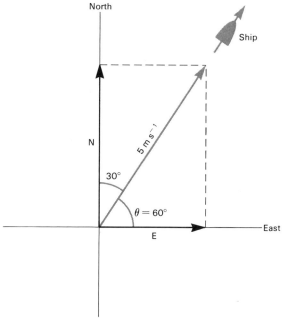

Fig. 1.3    Information for Worked Example 3

(b)    From equation [1.5]

$$V = T \sin\theta$$

with          $V = N$,    $T = 5\,\text{m s}^{-1}$,    $\theta = 60°$. So

$$N = 5 \sin 60° = 4.33$$

The ship travels northerly at $4.33\,\text{m s}^{-1}$. The time taken to travel 3 km North is given by

$$t = \frac{3000}{4.33} = 693$$

The ship takes 693 s to travel 3 km towards North. Note that the ship takes less time to travel in a northerly direction because its component of velocity is greater in this direction.

## WORKED EXAMPLE 4

A shell is fired at $400\,\text{m s}^{-1}$ at an angle of $30°$ to the horizontal. If the shell stays in the air for 40 seconds, calculate how far it lands from its original position. Assume that the ground is horizontal and that air resistance may be neglected.

*Solution*

Fig. 1.4 shows the trajectory of the shell. We require the range $R$.

From equation [1.4], the horizontal component $H$ is

$$H = T \cos\theta$$

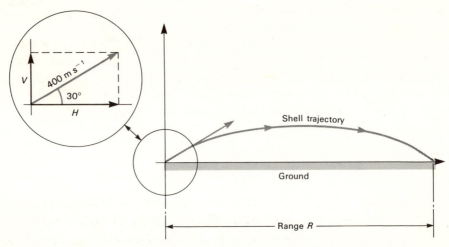

Fig. 1.4    Information for Worked Example 4

where                $T = 400\,\text{m s}^{-1}$   and   $\theta = 30°$. So

$H = 400 \cos 30° = 347$

The initial horizontal component of velocity is $347\,\text{m s}^{-1}$. This remains unchanged in the absence of air resistance. The range $R$ is given by

$R = \text{Horizontal component of velocity} \times \text{Time}$

$= 347 \times 40$

$= 13\,880$

The shell lands 13.88 km from its original position.

Note that the vertical component of velocity changes since it is subject to gravitational effects. This determines the time for which the shell stays in the air and explains the shape of the shells flight shown in Fig. 1.4.

## NEWTON'S LAWS OF MOTION

These are statements concerning the way bodies behave when subjected to systems of forces. They are based on acute observation and experimentation. All our experience of natural phenomena tends to confirm that they are correct.

## NEWTON'S FIRST LAW

A body will remain at rest or will continue to move with uniform motion in a straight line unless acted upon by a resultant external force.

## RESULTANT EXTERNAL FORCE

This will be either a single force acting alone or the resultant of several forces acting simultaneously. The force is one that is applied *to* the body from some source outside the body. This resultant force causes the body to accelerate. In the absence of a resultant force there is no acceleration so the body remains at rest or moves with constant velocity.

## INERTIA

The reason why a body behaves in accordance with Newton's first law is because of *inertia*. Inertia is a property which all bodies possess causing them to resist a change of motion and, therefore, enabling them to stay at rest or, if already moving, to go on moving uniformly in a straight line without the application of a force.

## MOMENTUM

It is easier to set in motion a small boat than a large tanker. Similarly far greater distances are required in which to bring to rest a tanker than are needed for a small boat. The massive tanker is said to possess more *momentum* than the boat. Similarly, larger distances are needed to stop an object if it is initially moving at high velocity than when moving with low velocity. So a fast-moving object has more momentum than a slow-moving object.

Momentum is defined as the product of mass and velocity

That is,

$$Momentum = Mass \times Velocity = mv$$

When two bodies of different mass are acted upon by the same force for the same length of time, the lighter one will attain a higher velocity than the heavier one but the momentum that each gains will be the same.

## FORCE AND MOMENTUM

The relationship between force and momentum is expressed by Newton's second law.

## NEWTON'S SECOND LAW

The rate of change of momentum of a body is proportional to the applied force and takes place in the direction in which the force acts.

If a resultant force of $F$ units acts on a body of mass $m$ kilograms for a time of $t$ seconds so that the velocity of the body changes uniformly from $u$ metres per second to $v$ metres per second,

$$\text{Rate of change of momentum} = \frac{\text{Change of momentum}}{\text{Time taken}}$$

$$= \frac{mv - mu}{t}$$

$$= \frac{m(v - u)}{t}$$

But
$$\frac{v - u}{t} = a$$

So that    Rate of change of momentum $= ma$

and according to Newton's second law, force is proportional to rate of change of momentum. Then

$$F \propto ma \quad \text{or} \quad F = kma \quad \text{where } k \text{ is a constant}$$

## THE UNIT OF FORCE

The equation $F = kma$ enables the unit of force to be defined in SI. The unit of force is chosen as that force which will give to a mass of 1 kg an acceleration of 1 metre per second per second. Upon substituting in the equation $F = kma$ we have $1 = k \times 1 \times 1$, so that $k = 1$. This unit of force is called the newton (N) hence when $F$ is in newtons, $m$ is in kilograms and $a$ is in metres per second per second.

$$F = ma \tag{1.6}$$

Note we have assumed that the mass $m$ of the body remains constant.

## MASS AND INERTIA

The mass of a body is usually defined as the quantity of matter in the body. All bodies possess inertia as a consequence of their mass and a force is necessary to overcome the inertia of the body, that is to accelerate it. Thus the greater the force needed to be applied to give a body a particular acceleration, the more massive that body is considered to be. The mass of a body determined in this way is sometimes referred to as the *inertial mass* of the body.

### WORKED EXAMPLE 5

Two bodies A and B are each given an acceleration of $8\,\mathrm{m\,s^{-2}}$. To do this a force of 24 N had to be applied to A and a force of 72 N to B. What is the mass of each body?

*Solution*

From equation [1.6],

$$F = ma$$

Therefore

$$\text{Mass of A} = \frac{\text{Force to overcome inertia of A}}{\text{Acceleration of A}}$$

$$= \frac{24}{8} = 3\,\mathrm{kg}$$

Similarly

$$\text{Mass of B} = \frac{\text{Force to overcome inertia of B}}{\text{Acceleration of B}}$$

$$= \frac{72}{8}$$

$$= 9\,\mathrm{kg}$$

### WORKED EXAMPLE 6

A resultant force of 25 N acts on a mass of 10 kg initially at rest. Find: (a) the acceleration, (b) the velocity reached after 6 seconds, (c) the distance travelled in this time.

*Solution*

(a)   From equation [1.6]

$$a = \frac{F}{m} = \frac{25}{10} = 2.5\,\mathrm{m\,s^{-2}}$$

(b)   Every second its velocity increases by $2.5\,\mathrm{m\,s^{-1}}$. After 6 s its velocity will be $6 \times 2.5 = 15\,\mathrm{m\,s^{-1}}$.

(c)   We have $u = 0$, $v = 15\,\mathrm{m\,s^{-1}}$ and $t = 6\,\mathrm{s}$.

$$\text{Distance travelled} = \text{Average velocity} \times \text{Time}$$

$$= \tfrac{1}{2}(u + v) \times t$$

$$= \tfrac{1}{2}(0 + 15) \times 6$$

$$= 45\,\mathrm{m}$$

## THE WEIGHT OF A BODY

The weight of a body is the force acting on it due to the pull exerted on it by the earth's gravitational attraction. When bodies are allowed to fall freely in a vacuum near to the earth's surface they all do so with an acceleration of $9.8 \, \mathrm{m \, s^{-2}}$ caused by this gravitational attraction. Applying equation [1.6] gives

Gravitational force or 'weight' of body  =  Mass of body × Acceleration

$$W = m \times 9.8$$

The acceleration of $9.8 \, \mathrm{m \, s^{-2}}$ caused by the earth's pull is denoted by the symbol $g$, so that it is usual to write

$$W = mg \qquad\qquad\qquad [1.7]$$

## ACTION AND REACTION

When a push is applied to an immovable object, say a wall, the wall reacts with an equal and opposite push. Similarly if the push is applied to a body which is free to move, this body will exert an equal and opposite push in the form of an inertia reaction as it is accelerated by the force. The same applies if pushes are replaced by pulls.

Newton embodied these facts in his third law as follows:

## NEWTON'S THIRD LAW

To every action there is an equal and opposite reaction.

### WORKED EXAMPLE 7

A body has a weight of 562 N on earth. (a) What force is required to give it an acceleration of $8 \, \mathrm{m \, s^{-2}}$? (b) What will be the reaction of the body when given this acceleration?

*Solution*

(a)  From equation [1.7]

$$W = mg$$

Therefore

$$\text{Mass of body} = m = \frac{W}{g} = \frac{562}{9.8} = 57.3 \, \mathrm{kg}$$

From equation [1.6]

$$F = ma$$

Therefore

$$\text{Accelerating force } F = 57.3 \times 8 = 458 \, \text{N}$$

(b)    The reaction of the body will, in accordance with Newton's third law, equal the force applied to the body and therefore is equal to 458 N.

## *WORKED EXAMPLE 8*

A car of mass 1200 kg is on a horizontal and slippery road. The road wheels slip when the push of the wheels on the road exceeds 500 N. Calculate the maximum acceleration of the car.

*Solution*

From equation [1.6]

$$a = \frac{F}{m} = \frac{500}{1200} = 0.417 \, \text{m s}^{-2}$$

The maximum acceleration of the car is 0.417 m s$^{-2}$. Note that it is the push of the road on the car wheels which causes the car to accelerate. This is equal and opposite to the push of the wheels on the road. If the road surface were frictionless, the car could not accelerate since the push of the road on the wheels would be zero. The wheels would 'spin'.

## *WORKED EXAMPLE 9*

Refer to Fig. 1.5. A car A of mass 1200 kg tows a caravan B of mass 800 kg. The two have an acceleration of 1.5 m s$^{-2}$. Assuming the only external force on the system is that between the driving wheels and the road, find: (a) the value of this force, (b) the tension in the coupling between car and caravan.

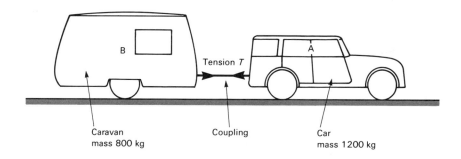

Caravan
mass 800 kg

Coupling

Car
mass 1200 kg

Fig. 1.5    Information for Worked Example 9

*Solution*

(a)    Apply equation [1.6] to car and caravan as a whole. With

$$m = \text{Total mass} = 1200 + 800 = 2000 \text{ kg}$$
$$F = ma = 2000 \times 1.5 = 3000 \text{ N}$$

This is the total resultant force which arises as a result of the interaction between driving wheels and road.

Applying equation [1.6] to the car alone gives

$$F = ma = 1200 \times 1.5 = 1800 \text{ N}$$

A net force of 1800 N acts on the car.

For the caravan alone

$$F = ma = 800 \times 1.5 = 1200 \text{ N}$$

A net force of 1200 N acts on the caravan.

(b)    Since a net force of 1200 N acts on the caravan, the tension *T* in the coupling is 1200 N.

Note that the net force on the car is 1800 N. This equals the total force of 3000 N *less* the tension in the coupling of 1200 N. This is another example of Newton's third law. Body A exerts a force on body B to the right. Body B exerts an equal force on body A to the left. In this case the force interchange is the tension *T* in the coupling, as shown in Fig. 1.5.

### WORKED EXAMPLE 10

A lift ascends with a uniform acceleration of $3 \text{ m s}^{-2}$ for a few seconds. It then rises at a uniform speed for a further time before being brought to rest with a retardation of $4 \text{ m s}^{-2}$. Find the force in the hoisting cable when: (a) the lift is accelerating, (b) when the lift is moving with uniform speed, and (c) when the lift is decelerating. The weight of the lift is 5 kN.

*Solution*

Let *T* be the force in the hoisting cable.

(a)    When the lift is accelerating upwards *T* must be greater than *W*, the weight of the lift (Fig. 1.6). Hence the resultant force causing acceleration is $T-W$. From equation [1.6]

$$T - W = ma$$

From equation [1.7]

$$m = \frac{W}{9.8} = \frac{5000}{9.8} = 510 \text{ kg}$$
$$T - 5000 = 510 \times 3$$
$$T = 5000 + 1530 = 6530 \text{ N} = 6.53 \text{ kN}$$

Hence when the lift is accelerating upwards the force in the hoisting cable is 6.53 kN.

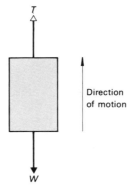

Fig. 1.6    The solution to Worked Example 10

(b)    When the lift is moving with uniform speed the force in the hoisting cable is equal to the weight of the lift, i.e. it is 5 kN.

(c)    When the lift is retarding the acceleration may be regarded as being negative. Hence

$$T - 5000 = -510 \times 4$$
$$T - 5000 = -2040$$

Therefore                 $T = 2960\,N = 2.96\,kN$

Hence when the lift is retarding the force in the hoisting cable is only 2.96 kN.

## GRAVITATIONAL FIELDS

Previously we stated that the weight $W$ of a body is the force acting on it due to the Earth's gravitational attraction. Gravitational forces occur since masses attract each other, solely as a result of the matter contained in each body. The force of attraction is very small unless one or both of the masses is very large. The Earth is a very massive body, and it attracts an object with a force which we refer to as its weight. Since the attraction force is directly proportional to the mass of the object then bodies allowed to fall freely, in the absence of air resistance, near to the Earth's surface will all do so with an acceleration of $9.8\,m\,s^{-2}$.

An alternative way of dealing with gravitational forces is to refer to the existence of a *gravitational field*. This is a region in which a body experiences a force because it is in the vicinity of a very massive object, such as planet Earth. We define

$$\text{Gravitational field strength} = \frac{\text{Force due to gravitational attraction}}{\text{Mass of body}} \quad [1.8]$$

From equation [1.7]

$$\text{Force due to gravitational attraction} = \text{Weight } W = mg$$

Thus from equation [1.8]

$$\text{Gravitational field strength} = \frac{mg}{m} = g$$

The gravitational field strength is thus equivalent to the acceleration due to gravity. Note that the units of gravitational field strength are $N\,kg^{-1}$ which are equivalent to the units of acceleration.

## WORKED EXAMPLE 11

Calculate the gravitational force on a man of mass 70 kg situated on the following planets:

(a)   Earth, gravitational field strength $= 9.8\,N\,kg^{-1}$

(b)   Mars, gravitational field strength $= 3.7\,N\,kg^{-1}$

(c)   Jupiter, gravitational field strength $= 25.0\,N\,kg^{-1}$

*Solution*

(a)   Gravitational force $= mg = 70 \times 9.8 = 686\,N$

(b)   Gravitational force $= mg = 70 \times 3.7 = 259\,N$

(c)   Gravitational force $= mg = 70 \times 25.0 = 1750\,N$

A man on Jupiter would have difficulty in supporting his own weight. On Mars he would be able to carry large masses.

**EXERCISE 1**

1)   A car accelerates from rest at $2\,m\,s^{-2}$ during a period of 5 seconds. Find the distance travelled during this acceleration.

2)   A train accelerates from rest at $0.2\,m\,s^{-2}$. Find the time taken and the distance travelled for the speed to reach $100\,km\,h^{-1}$.

3)   A road vehicle at a given instant has a speed of $6\,m\,s^{-1}$. It is given a uniform retardation of $0.9\,m\,s^{-2}$ until its speed is reduced to $3\,m\,s^{-1}$. Find: (a) the time taken in reducing speed, (b) the distance travelled during retardation.

4)   A vehicle is accelerated at $1.5\,m\,s^{-2}$ from a speed of $8\,m\,s^{-1}$. Calculate: (a) the time required to attain a speed of $25\,m\,s^{-1}$, (b) the distance travelled in this time.

5)   A lift, starting from rest, increases speed uniformly for the first 10 s until a speed of $3 \, \text{m s}^{-1}$ is reached. This speed then remains constant for the next 6 s and finally the lift is brought to rest in a further 8 s of uniform retardation. Determine: (a) the initial acceleration, (b) the final retardation, (c) the height travelled.

6)   Explain how velocity differs from speed.

7)   John sets off from the corner of a rectangular field with a velocity of $2 \, \text{m s}^{-1}$ at an angle of $40°$ to the shorter side of the field which is 200 m by 500 m. How long does he take to reach a boundary? How far up the longer boundary will he be?

8)   A ship moves with velocity $8 \, \text{m s}^{-1}$ in a direction N20°W. Calculate its northerly component of velocity and the time taken to travel 20 km North.

9)   A shell is fired from a cannon with a velocity of $300 \, \text{m s}^{-1}$ at $25°$ to the horizontal. Find the horizontal range of the shell if it stays in the air for 54 seconds.

10)   A cannon fires shells at an angle of $30°$ to the horizontal and from the top of a vertical cliff. The shells hit the sea at a horizontal distance of 4.5 km from the foot of the cliff. If the shells stay in the air for 25 s, find the velocity of the shells on leaving the cannon.

11)   A body of mass 50 kg is free to move. If a force of 5 newtons is applied what will be the acceleration given to the body?

12)   The force causing an iron core to move in an electric solenoid is 2 N and the mass of the core is 300 grams. With what acceleration will it move?

13)   A force of 20 N acts on a mass of 5 kg initially at rest and accelerates it to a final velocity of $16 \, \text{m s}^{-1}$. Calculate: (a) the acceleration, (b) the distance travelled during the acceleration.

14)   A force of 10 N acts for 14 s on a mass of 50 kg initially at rest. Find the velocity acquired.

15)   A motorist has a personal reaction time of 0.5 s. He drives a car with total mass, including himself, of 1500 kg. If his average braking force is 12 000 N, calculate the minimum distance he can stop in when travelling at $72 \, \text{km h}^{-1}$ $(20 \, \text{m s}^{-1})$.

16)   An aircraft of mass 20 000 kg lands on an aircraft carrier deck with a speed of $100 \, \text{m s}^{-1}$. Calculate the value of the force required to bring it to rest in a distance of 100 m.

**17)**  A body having a mass of 20 kg falls freely from rest near to the Earth's surface. (a) What will be the gravitational force on the body? (Assume $g = 9.8 \, \mathrm{m\,s^{-2}}$.)  (b) How far will the body fall during a period of 5 seconds?

**18)**  A body which is free to move is given an acceleration of $72 \, \mathrm{m\,s^{-2}}$. Because of this acceleration and the mass of the body its inertial reaction amounts to 3600 N. Calculate the mass of the body.

**19)**  A hoist and its load have a total weight of 4 kN. They are lifted vertically reaching a velocity of $6 \, \mathrm{m\,s^{-1}}$ after rising 15 m from rest. Assuming that the acceleration is uniform, calculate the net upward force causing acceleration.

**20)**  A body having a mass of 2.5 kg is moving with a velocity of $4 \, \mathrm{m\,s^{-1}}$. A uniform force acts on it and causes its velocity to increase to $12 \, \mathrm{m\,s^{-1}}$ in 4 s. Calculate the magnitude of the force.

**21)**  A load of 3 kN is raised by a chain and given a starting acceleration of $0.5 \, \mathrm{m\,s^{-2}}$. Determine the total initial pull in the chain.

**22)**  A lorry weighing 20 000 N is moving at $50 \, \mathrm{km\,h^{-1}}$. It is brought to rest in a distance of 28 m. Find the average braking force exerted.

**23)**  A train of mass 15 000 kg pulls a truck of mass 2000 kg and starts with an acceleration of $0.2 \, \mathrm{m\,s^{-2}}$. Find: (a) the tension in the coupling between truck and train; (b) the net force acting on the train.

**24)**  In Question 23 another truck of mass 3000 kg is coupled to the first truck. Assuming that the only external force on the system is between the train driving wheels and track, calculate: (a) the value of this force; (b) the tension in the coupling between train and truck 1, (c) the tension in the coupling between truck 1 and truck 2.

**25)**  A man weighing 560 N stands in a lift which moves with a uniform acceleration of $4 \, \mathrm{m\,s^{-2}}$. Find the reaction between the man and the floor of the lift when the lift is: (a) ascending, (b) descending.

**26)**  A lift cage weighing 30 kN is suspended from a wire rope. Determine the motion of the cage when the tension in the rope is: (a) 28 kN, (b) 30 kN, and (c) 32 kN. Assume that the cage is ascending.

# 2 CIRCULAR MOTION

## BASIC DEFINITION

Many machines contain components which rotate about an axis. A centrifuge and a lathe are examples. In this chapter we shall study the motion of a body moving with *uniform speed* at a *fixed distance* from a *fixed axis*. This is called *uniform circular motion.*

## ANGULAR DISPLACEMENT

For rotational motion we measure the displacement in angular terms (radians) rather than linear terms (metres).

Refer to Fig. 2.1. The object moves with uniform speed $v$ around the circumference of the circle of centre O. The angular displacement $\theta$ is the angle swept out by the line joining the body to the centre of the circle.

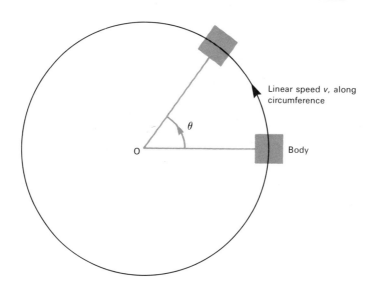

Fig. 2.1   Circular motion — angular displacement

## RADIAN MEASURE

Angular displacement is usually measured in radians. Fig. 2.2 explains radian measure. Fig. 2.2(a) shows that one radian is the angle subtended at the centre of a circle by an arc whose length is equal to the radius of the circle. Fig. 2.2(b) shows that, in general, $\theta = \dfrac{s}{r}$.

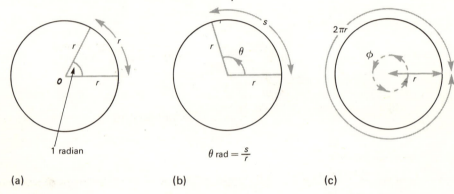

(a)                              (b)                              (c)

Fig. 2.2    Radian measure

$$\theta \,(\text{rad}) = \frac{\text{Length of arc}}{\text{Radius}} = \frac{s}{r} \qquad [2.1]$$

Fig. 2.2(c) shows that, for one complete revolution,

$$\theta = \frac{s}{r} = \frac{2\pi r}{r} = 2\pi$$

So there are $2\pi$ radians in 1 revolution. So

$$2\pi \text{ rads} = 360 \text{ degrees}$$

$$\therefore \qquad 1 \text{ rad} = \frac{360}{2\pi} = 57°18'$$

## ANGULAR VELOCITY

Angular velocity is the angle turned through per second. That is

$$\text{Angular velocity} = \frac{\text{Angle turned through (rad)}}{\text{Time taken (s)}}$$

or, in symbols,

$$\omega = \frac{\theta}{t} \qquad [2.2]$$

where    $\omega$ = angular velocity in radians per second (rad s$^{-1}$)
    $\theta$ = angle turned through in radians (rad)
    $t$ = time taken in seconds (s)

Frequently the angular speed is stated in revolutions per minute (rev min$^{-1}$ or rpm). It is then necessary to convert the speed to radians per second for the purpose of calculation.

$$1 \text{ revolution} = 2\pi \text{ radians}$$

$$1 \text{ revolution per minute} = \frac{2\pi}{60} \text{ radians per second}$$

If the body rotates at $N$ rev min$^{-1}$,

$$\omega = \frac{2\pi N}{60} \text{ radians per second} \qquad [2.3]$$

### WORKED EXAMPLE 1

The armature of an electric motor rotates at 5000 rev min$^{-1}$. Calculate its angular speed in radians per second.

### Solution

Since $N = 5000$ rev min$^{-1}$, from equation [2.3]

$$\omega = \frac{2 \times \pi \times 5000}{60} = 524 \text{ rad s}^{-1}$$

## THE RELATION BETWEEN ANGULAR AND LINEAR SPEEDS —

Let a body move in a circular path of radius $r$ metres. In a time of $t$ seconds let it move through an arc of length $s$ metres corresponding to an angle of $\theta$ radians (Fig. 2.3).

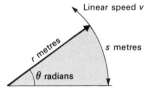

Fig. 2.3    Radial motion through $s$ metres

Let the linear speed of the body along the circumference of the circle be $v$.

Rearranging equation [2.1] gives

$$s = r\theta$$

Dividing both sides of this equation by $t$ gives

$$\frac{s}{t} = \frac{r\theta}{t}$$

But linear speed $v = \dfrac{s}{t}$ and $\omega = \dfrac{\theta}{t}$ (equation [2.2]), so

$$v = \omega r \qquad\qquad [2.4]$$

Equation [2.4] therefore relates angular and linear speeds.

### WORKED EXAMPLE 2

A pulley has a diameter of 300 mm and it rotates at 240 r.p.m. It drives a belt which passes round the rim. Calculate the speed of the belt in metres per second.

### Solution

We are given $N = 240$ rev min$^{-1}$. From equation [2.3],

$$\omega = \frac{2\pi N}{60} = \frac{2 \times \pi \times 240}{60} = 25.1 \text{ rad s}^{-1}$$

The radius of the pulley is $r = 150$ mm $= 0.15$ m. Using equation [2.4] gives

$$v = \omega r = 25.1 \times 0.15 = 3.77 \text{ m s}^{-1}$$

Hence the speed of the belt is $3.77$ m s$^{-1}$.


## ACCELERATION IN A CIRCLE

Fig. 2.4 shows a body attached to a string and moving in a circular path with centre O. Although the body has a constant speed its velocity is changing because its direction of motion is changing. The body is therefore accelerating at all times. Thus there must be an external force acting on the body. In this case the force comes from the tension $T$ in the string. Since the tension acts on the body towards the centre O of the circle, then the acceleration is directed to the centre of the circle.

Suppose the body moves with uniform speed $v$ in a circle of radius $r$ (Fig. 2.4). We show below that the acceleration towards the centre is given by

$$a = \frac{v^2}{r} \qquad\qquad [2.5]$$

This is called the *centripetal acceleration*.

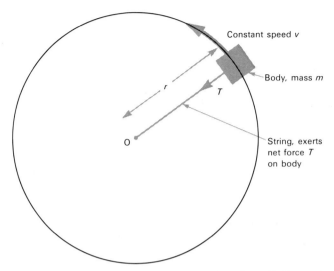

Constant speed *v*

Body, mass *m*

*T*

String, exerts
net force *T*
on body

*O*

*r*

Fig. 2.4   Uniform circular motion

The force $F$ required to keep a body of mass $m$ moving with speed $v$ in a circle of radius $r$ is, from equation [1.6] and [2.5], given by

$$F = ma = \frac{mv^2}{r} \qquad [2.6]$$

This is termed the *centripetal force.*

When a satellite orbits the earth the centripetal force is provided by gravitational attraction forces. For electrons in orbit around a nucleus it is provided by electrostatic forces.

### WORKED EXAMPLE 3

An object of mass 3 kg moves in a circle of radius 5 m with a constant speed of 8 m s$^{-1}$. Calculate: (a) the centripetal acceleration, (b) the force needed toward the centre.

*Solution*

(a)   From equation [2.5]

$$a = \frac{v^2}{r}$$

We have $v = 8\,\text{m s}^{-1}$ and $r = 5\,\text{m}$. So

$$a = \frac{8^2}{5} = 12.8$$

The acceleration of the mass is 12.8 m s$^{-2}$ toward the centre of the circle.

(b)    From equation [2.6] with $m = 3\,\text{kg}$,

$$F = m\frac{v^2}{r} = 3 \times 12.8 = 38.4\,\text{N}$$

An inwards force of 38.4 N is necessary to keep the mass moving in a circle.

## WORKED EXAMPLE 4

A spaceman in training is revolved in a horizontal circle of radius 6 m. If the maximum acceleration he is to withstand is $70\,\text{m}\,\text{s}^{-2}$ calculate: (a) his speed of rotation, (b) the rate of rotation in r.p.m.

*Solution*

(a)    Rearranging equation [2.5] gives

$$v^2 = ra$$

where   $r = 6\,\text{m}$ and $a = 70\,\text{m}\,\text{s}^{-2}$

$$v = \sqrt{6 \times 70} = 20.5$$

The speed of rotation is $20.5\,\text{m}\,\text{s}^{-1}$.

(b)    In 1 second the spaceman travels $v$ metres. In 1 minute he travels $60\,v$ metres. Each revolution he travels $2\pi r$ metres. If he makes $N$ r.p.m., then

$$N = \frac{60v}{2\pi r} = \frac{60 \times 20.5}{2\pi \times 6} = 32.6$$

The spaceman makes 32.6 r.p.m.

# THE CENTRIFUGE

Small particles in suspension in a liquid take a long time to settle out. If this mixture is placed in a tube in a centrifuge and spun at high speed, the particles settle out much more quickly. This is because the forces acting on the particle are much larger when it is in circular motion at high speed.

## WORKED EXAMPLE 5

(a)    A spherical particle of radius $1\,\mu\text{m}$ has a density of $1200\,\text{kg}\,\text{m}^{-3}$. Find the resultant force on it when it is stationary and suspended in still water (density $1000\,\text{kg}\,\text{m}^{-3}$).

(b)    The mixture containing the particle is now spun round in a centrifuge, in a horizontal plane, with a nominal radius of 0.30 m and at 60 revolutions per second. Find the force acting on the particle.

*Solution*

(a)    Refer to Fig. 2.5. The resultant force $F$ on the (initially) stationary particle is downwards and is equal to its weight less the Archimedean upthrust due to the water. From equation [1.7], the weight $W = mg$, so

$$F = mg - m'g = (m - m')g$$

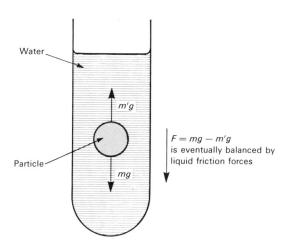

Fig. 2.5    Particle motion under gravity

where   $m$ = Mass of particle and
                $m'$ = Mass of an equal volume of water

The particle is spherical. So

$$m = \tfrac{4}{3}\pi r^3 \rho \quad \text{and} \quad m' = \tfrac{4}{3}\pi r^3 \rho'$$

where   $r$ = radius of particle = $1 \times 10^{-6}$ m
                $\rho$ = Density of particle = 1200 kg m$^{-3}$
                $\rho'$ = Density of water = 1000 kg m$^{-3}$

Substituting these values gives

$$m = \tfrac{4}{3}\pi \times (1 \times 10^{-6})^3 \times 1200 = 5.03 \times 10^{-15} \text{ kg}$$
$$m' = \tfrac{4}{3}\pi \times (1 \times 10^{-6})^3 \times 1000 = 4.19 \times 10^{-15} \text{ kg}$$

So, since $g = 9.8$ m s$^{-2}$,

$$F = (m - m')g = (0.84) \times 10^{-15} \times 9.8$$
$$= 8.24 \times 10^{-15} \text{ N}$$

A net force of $8.24 \times 10^{-15}$ N acts on a stationary particle in a still mixture. This force causes the particle to fall through the liquid. The forces of liquid friction *oppose* the motion of the particle. When liquid friction forces are equal and opposite to $F$ the particle falls with constant velocity. The velocity is very small in this case.

(b)   Since the particle is in a liquid then, due to the Archimedean effect, it acts like a particle of effective mass $(m - m')$.

Suppose the mixture is now spun in a circle of radius $R$ and with speed $v$ (see Fig. 2.6). The centripetal force $F'$ required to keep the particle in circular motion is, from equation [2.6], given by

$$F' = (m - m') \frac{v^2}{R}$$

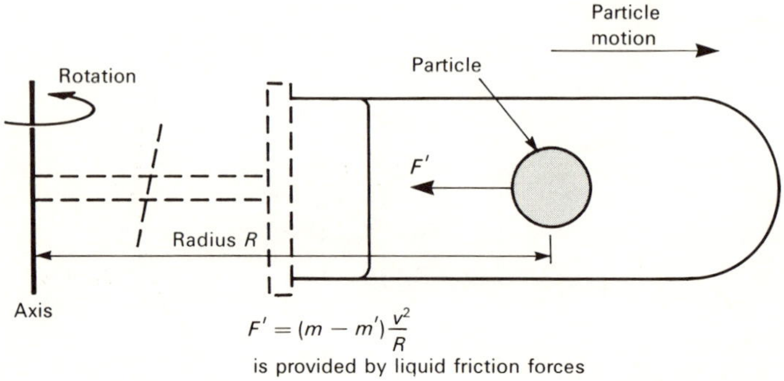

$$F' = (m - m') \frac{v^2}{R}$$
is provided by liquid friction forces

Fig. 2.6   Particle motion in a centrifuge

We have    $m = 5.03 \times 10^{-15}$ kg  and  $m' = 4.19 \times 10^{-15}$ kg

$R = $ Radius of circle $= 0.30$ m

$v = $ Linear speed

= Circumference of circle $\times$ Number of revolutions per second

$= (2\pi \times 0.3) \times 60$

$= 113 \, \text{m s}^{-1}$

Substituting gives

$$F' = \frac{(0.84 \times 10^{-15}) \times (113)^2}{0.3}$$

$$= 3.58 \times 10^{-11} \, \text{N}$$

An initial net force of $3.58 \times 10^{-11}$ N acts on the particle in the centrifuge. Note from (a) and (b) that $F'$ is very much greater than $F$. This means the particle settles out much more quickly under the action of a centrifuge. Using the equations of fluid mechanics, we can show that for case (a), fall under gravity, the particle takes $25 \times 10^3$ s to fall 1 cm. In case (b), using the centrifuge, the particle takes only 5.8 s to travel 1 cm.

## THE PROOF OF THE EQUATION FOR CENTRIPETAL ACCELERATION

Refer to Fig. 2.7(a). Suppose a particle is moving in uniform circular motion with linear speed $v$ along the circumference of the circle of radius $r$. Suppose the particle moves from A to B in a small time $\delta t$. Its angular displacement is $\delta\theta$. The speed of the particle is the same at A and B but its velocity is different since the direction of motion has changed.

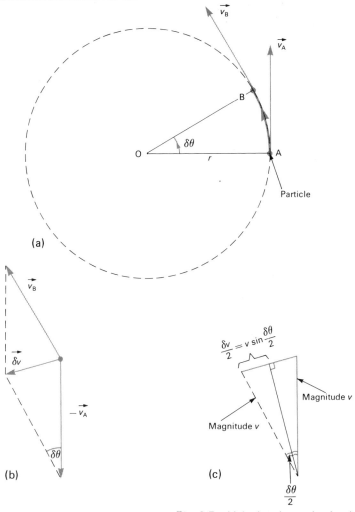

Fig. 2.7    Velocity change in circular motion

Let

$$\vec{v_A} = \text{Velocity at A}$$

$$\vec{v_B} = \text{Velocity at B}$$

The arrows emphasise the directional nature of velocity. So velocity change $\vec{\delta v}$ as the particle moves from A to B is given by

$$\vec{\delta v} = \vec{v_B} - \vec{v_A} = \vec{v_B} + (-\vec{v_A})$$

Fig. 2.7(b) shows how the vector quantity $\vec{v_B} + (-\vec{v_A})$ is found using the parallelogram rule.

From trigonometry (see Fig. 2.7(c)) the magnitude of $\vec{\delta v}$ is given by

$$\delta v = 2v \sin \frac{\delta\theta}{2} = v\delta\theta$$

since when $\delta\theta$ is small, $\sin \frac{\delta\theta}{2} = \frac{\delta\theta}{2}$.

Now acceleration $a$ is given by

$$a = \frac{\text{Change in velocity}}{\text{Time}} = \frac{\delta v}{\delta t}$$

But $\delta v = v\delta\theta$, so

$$a = v\frac{\delta\theta}{\delta t} \qquad\qquad [2.7]$$

From equation [2.2] we see $\frac{\delta\theta}{\delta t} = \omega$. Also from equation [2.4] we see $\omega = \frac{v}{r}$. So substituting into equation [2.7] gives

$$a = v\frac{\delta\theta}{\delta t} = v\omega = \frac{v^2}{r}$$

The centripetal acceleration is of magnitude $\frac{v^2}{r}$. Note from Fig. 2.7(b) that the velocity change $\vec{\delta v}$ is directed toward the centre. Thus the acceleration is toward the centre.

Note also from equation [2.4] that since $v = r\omega$, then

$$a = \frac{v^2}{r} = \frac{(r\omega)^2}{r} = r\omega^2$$

where $\omega$ is the angular speed of the particle motion.

## EXERCISE 2

1)   A turbine rotor revolves at $628 \text{ rad s}^{-1}$. What is the speed of the rotor in revolutions per minute? (Assume $\pi = 3.14$.)

2)   A pulley having an effective diameter of $300 \text{ mm}$ is driven by a vee-belt at $360$ r.p.m. Calculate the linear speed of the belt correct to two places of decimals (Assume $\pi = 3.14$.)

**3)**   A pulley has a diameter of 0.4 m and it rotates at 300 r.p.m. It drives a belt which passes round the rim. Calculate the speed of the belt in metres per second. (Assume $\pi = 3.14$.)

**4)**   An object of mass 5 kg moves in a circle of radius 4 m. If the uniform speed of the object is $6\,\mathrm{m\,s^{-1}}$, find: (a) the centripetal acceleration; (b) the force needed toward the centre.

**5)**   A car of mass 800 kg is moving at $40\,\mathrm{m\,s^{-1}}$ around a bend of radius 0.5 km. What centripetal force is required to keep the car moving around the bend and where does this come from?

**6)**   An object of mass 6 kg is whirled around in a *vertical* circle of radius 2 m with a speed of $10\,\mathrm{m\,s^{-1}}$. Calculate the maximum and minimum tension in the string connecting the object to the centre of the circle. (Assume $g = 9.8\,\mathrm{m\,s^{-2}}$.)

**7)**   A car travels over a humpback bridge with a radius of curvature of 30 m. Calculate the maximum speed of the car if its road wheels are to stay in contact with the bridge. (Assume $g = 9.8\,\mathrm{m\,s^{-2}}$.)

**8)**   In a simple model of the hydrogen atom the electron is considered to move in a circle of radius $5 \times 10^{-11}\,\mathrm{m}$. It circles the nucleus $7 \times 10^{15}$ times each second. Calculate for the electron: (a) its angular speed; (b) its linear speed; (c) the centripetal acceleration; (d) the centripetal force. (Mass of electron $= 9 \times 10^{-31}\,\mathrm{kg}$.)

**9)**   (a)   A spherical particle of radius $4\,\mu\mathrm{m}$ has a density of $1200\,\mathrm{kg\,m^{-3}}$ Find the resultant force on a stationary particle when suspended in still water (density, $1000\,\mathrm{kg\,m^{-3}}$).

     (b)   The mixture containing the particle is now spun round in a centrifuge, in a horizontal plane, with a nominal radius of 0.40 m and at 3000 revolutions per minute. Find the force acting on the particle.

# 3 LINEAR MOMENTUM

## MOMENTUM

In Chapter 1 we saw that the momentum of a body depends upon its mass — it requires far greater distances to bring to rest a tanker than a small boat. Momentum also depends upon velocity since the greater the velocity of, say, a tanker the greater the distance needed to bring it to rest.

Momentum is defined as the product of mass and velocity. That is,

$$\text{Momentum} = \text{Mass} \times \text{Velocity} = mv \qquad [3.1]$$

*WORKED EXAMPLE 1*

Fig. 3.1 shows a body A of mass 3 kg moving to the right with a velocity of 20 m s⁻¹. Body B is of mass 5 kg and moves to the left with velocity 8 m s⁻¹. Calculate: (a) the momentum of A, (b) the momentum of B, (c) the total momentum of A and B.

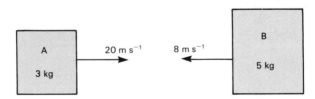

Fig. 3.1   Information for Worked Example 1

*Solution*

(a)                           Momentum of A $= mv$

$$= 3 \times 20$$

$$= 60 \, \text{kg m s}^{-1}$$

Note that the units of momentum are those of mass times velocity $= \text{kg} \times \text{m s}^{-1} = \text{kg m s}^{-1}$.

(b)   In Chapter 1 it was stated that velocity is a vector quantity. In part (a) we *assumed* that objects moving to the right have positive velocity. Thus objects moving to the left have negative velocity.

**Thus**

$$\text{Momentum of B} = mv$$
$$= 5 \times -8$$
$$= -40 \text{ kg m s}^{-1}$$

Note that since velocity is a vector quantity then momentum is also a vector quantity.

(c)   (Total momentum of A and B) = (Momentum of A) + (Momentum of B)

$$= 60 - 40$$
$$= 20 \text{ kg m s}^{-1}$$

## IMPULSE

In Chapter 1 it was shown that if a force of $F$ units acts on a body of mass $m$ kilograms for a time of $t$ seconds, so that the velocity of the body changes from $u$ metres per second to $v$ metres per second, then

$$\text{Rate of change of momentum} = \frac{mv - mu}{t}$$

According to Newton's second law, force is proportional to rate of change of momentum.

Hence

$$F \propto \frac{mv - mu}{t}$$

or

$$F = k\left(\frac{mv - mu}{t}\right)$$

where $k$ is a constant. If $F$ is in newtons, then $k = 1$. Hence

$$F = \frac{mv - mu}{t} \qquad [3.2]$$

That is,

$$\text{Force} = \text{Rate of change of momentum}$$

Rearranging equation [3.2] gives

$$F \times t = mv - mu$$
$$= \text{Change of momentum of body} \qquad [3.3]$$

The product of force and time is called the *impulse* of the force. That is,

$$\text{Impulse} = \text{Force} \times \text{Time for which force acts}$$

$$= F \times t$$

Equation [3.3] tells us

Impulse = Change of momentum of body

Note that the *change* in momentum due to an external force $F$ acting depends only upon $F$ and $t$. It does not depend upon the mass of the body.

Note, too, that in general impulses act for very short times.

*WORKED EXAMPLE 2*

A golf ball, initially at rest, is hit with a golf club which exerts an average force of 70 N over a time of 0.02 s.

Calculate: (a) the change in momentum of the ball, (b) the velocity it acquires if it has a mass of 0.02 kg.

*Solution*

(a)    From equation [3.3]

$$\text{Change of momentum} = F \times t$$

$$= 70 \times 0.02$$

$$= 1.4\,\text{N s}$$

Note that the unit N s is the same as kg m s$^{-1}$ since a newton has the basic unit kg m s$^{-2}$ (mass $\times$ acceleration).

(b)    Change of momentum $= mv - mu$.
Hence from part (a)

$$mv - mu = 1.4$$

But $m = 0.02$ and $u = 0$. Hence

$$0.02 \times v - 0 = 1.4$$

or

$$v = \frac{1.4}{0.02} = 70\,\text{m s}^{-1}$$

*WORKED EXAMPLE 3*

Coal is deposited at a uniform rate of 50 kg each second on to a conveyor belt moving horizontally at 0.4 m s$^{-1}$. Assuming that the coal has negligible speed just before it is placed on the belt, calculate the force required to pull the belt.

*Solution*

A force is needed to pull the belt because the momentum of the coal in the direction of the belt is being increased.

Force required = Rate of change of momentum

or
$$F = \frac{mv - mu}{t}$$
[3.2]

Now

Initial velocity of coal $u = 0$
Final velocity of coal $v = 0.4\,\mathrm{m\,s^{-1}}$

Also, in a time of 1 second a mass of 50 kg of coal is deposited on the belt. Hence

$t = 1$   for   $m = 50\,\mathrm{kg}$

From equation [3.2]

$$F = \frac{50 \times 0.4 - 0}{1} = 20\,\mathrm{N}$$

## THE PRINCIPLE OF CONSERVATION OF MOMENTUM _____

When two, or more, bodies collide we find that, provided that no *external* forces such as friction act, the total momentum of the bodies is the same before a collision as afterwards. The only forces which act are those *during* the collision when the bodies exert equal, and opposite, forces on each other. These forces are 'internal' forces and, as shown on p. 35, do not produce a change in the total momentum of the bodies.

Fig. 3.2(a) shows a body of mass $m_1$ and initial velocity $u_1$ which is about to overtake and collide with a body of mass $m_2$ moving with initial velocity $u_2$ along the same direction. After collision the respective velocities are $v_1$ and $v_2$ as shown in Fig. 3.2(b).

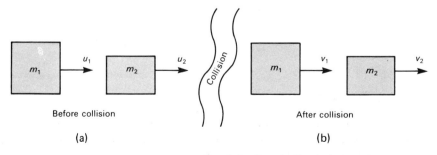

Fig. 3.2   Two bodies before and after collision

The principle of conservation of momentum states that:

If no external forces act on a system of colliding objects, then the total momentum of the objects remains constant.

Hence, referring to Fig. 3.2:

$$\left(\begin{array}{c}\text{Total momentum in a particular}\\ \text{direction before collision}\end{array}\right) = \left(\begin{array}{c}\text{Total momentum in the same}\\ \text{direction after collision}\end{array}\right)$$

$$m_1u_1 + m_2u_2 = m_1v_1 + m_2v_2 \qquad [3.4]$$

## WORKED EXAMPLE 4

A 2 kg mass moving with a velocity of 8 m s$^{-1}$ collides with a 3 kg mass moving with a velocity of 6 m s$^{-1}$. If the two masses stick together on impact calculate their common velocity when: (a) the two masses are initially moving in the same direction, (b) the two masses are initially moving in opposite directions.

*Solution*

(a)    Fig. 3.3 shows the bodies before and after collision. Since $v_1 = v_2 = v$, then, from equation [3.4],

$$m_1u_1 + m_2u_2 = m_1v + m_2v$$
$$= (m_1 + m_2)v$$

Thus

$$2 \times 8 + 3 \times 6 = 5v$$

$$v = \frac{34}{5} = 6.8 \, \text{m s}^{-1}$$

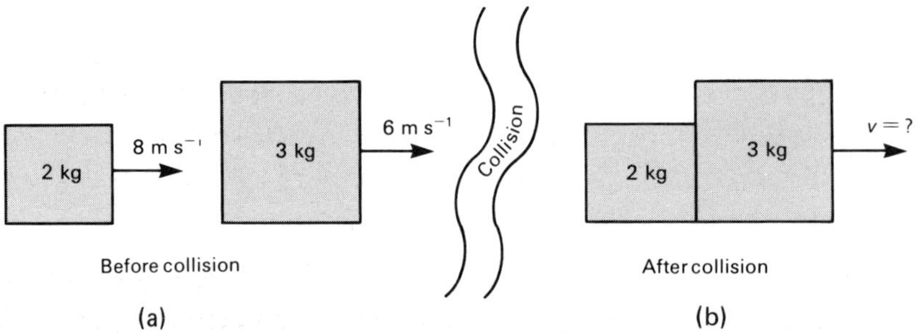

| | |
|---|---|
| Before collision | After collision |
| (a) | (b) |

Fig. 3.3    Information for Worked Example 4(a)

(b)    Fig. 3.4 shows the bodies before and after collision. As in Worked Example 1 we assume that bodies moving to the right have a positive velocity, so that bodies moving to the left have a negative velocity. Thus $u_1 = +8$ and $u_2 = -6$. Since $v_1 = v_2 = v$, then, from equation [3.4],

$$m_1u_1 + m_2u_2 = (m_1 + m_2)v$$
$$2 \times 8 - 3 \times 6 = 5v$$

$$\therefore \qquad v = -\frac{2}{5} = -0.4 \, \text{m s}^{-1}$$

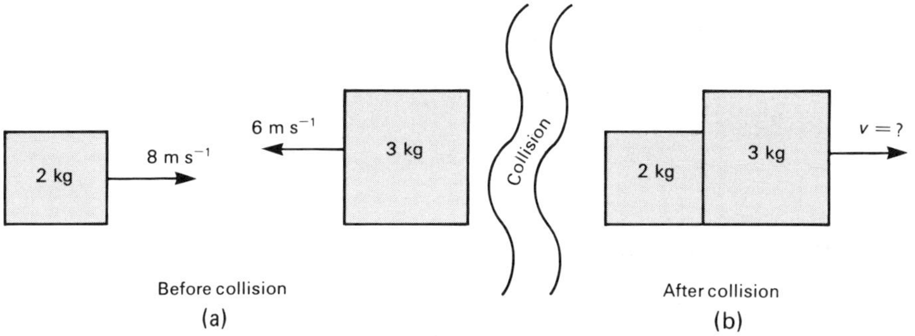

Fig. 3.4   Information for Worked Example 4(b)

The negative value of $v$ means that the combined masses move to the *left* after collision. This must be so since the momentum of the 3 kg mass is greater than the momentum of the 2 kg mass. Since the magnitude of the momentum of each mass is almost the same they almost cancel each other out, resulting in a small common velocity after collision. During this type of collision a large fraction of the initial kinetic energy* is dissipated.

When other bodies collide, particularly hard objects which separate after collision, little or no kinetic energy may be lost. This would be true for steel balls colliding, in which case only a little kinetic energy is lost.

## EXPLOSIONS

When an object, such as a space vehicle, splits into separate parts it does so as a result of *internal* forces. The total momentum of the separate parts is thus the same as that of the original body. The total kinetic energy of the fragments, may be more than that of the original body because of the work done by the internal forces during the 'explosion'.

A simple example is shown in Fig. 3.5(a). The initial object consists of two trolleys which have been pushed together, so compressing spring S. Energy is thus stored in the system. If the spring is now released, by tapping pin P, then the trolleys A and B move off in opposite directions, as in Fig. 3.5(b). Suppose the two trolleys were initially at rest, so their initial momentum is zero. Since linear momentum is conserved:

*Kinetic energy is explained in Chapter 4.

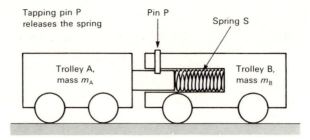

(a)  Spring compressed, trolleys at rest

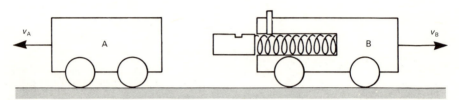

(b)  Spring released, trolleys in motion

Fig. 3.5    Two trolleys springing apart

$$0 = \text{Momentum of A} + \text{Momentum of B}$$

or      $$0 = m_A v_A \qquad + m_B v_B \qquad \qquad [3.5(a)]$$

or      $$v_A = -\left(\frac{m_B}{m_A}\right) v_B \qquad \qquad [3.5(b)]$$

The relationship between the speeds of the separate parts, as given by equations [3.5] can be investigated experimentally using trolleys and ticker timers, or an air track system. The negative sign in equation [3.5(b)] indicates that A moves in the opposite direction to B.

### WORKED EXAMPLE 5

Refer to Fig. 3.6 in which two trolleys are separated by a compressed spring. If the spring is now released and the 2 kg trolley moves with a velocity of 0.6 m s⁻¹, calculate the velocity of the 1.5 kg trolley. Neglect the mass of the spring.

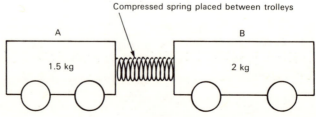

Fig. 3.6    Information for Worked Example 5

*Solution*

Assuming the two trolleys were initially at rest, then from equation [3.5(a)]

$$0 = m_A v_A + m_B v_B$$

where
$$m_A = 1.5 \, \text{kg}$$
$$m_B = 2.0 \, \text{kg}$$
$$v_B = +0.6 \, \text{m s}^{-1}$$

Then
$$0 = 1.5 \times v_A + 2.0 \times 0.6$$

or
$$v_A = \frac{-2.0 \times 0.6}{1.5}$$
$$= -0.8 \, \text{m s}^{-1}$$

## WORKED EXAMPLE 6

A shell of mass 0.8 kg is fired from a gun with a velocity of 250 m s⁻¹. If the gun has a mass of 500 kg, calculate the velocity of recoil of the gun.

*Solution*

We assume that the shell and gun were initially at rest. From equation [3.5(a)]

$$0 = m_A v_A + m_B v_B$$

where $m_A$ = mass of gun = 500 kg
$m_B$ = mass of shell = 0.8 kg
$v_B$ = velocity of shell = 250 m s⁻¹
$v_A$ = recoil velocity of gun

Then
$$0 = 500 \times v_A + 0.8 \times 250$$

or
$$v_A = \frac{-0.8 \times 250}{500}$$
$$= -0.4 \, \text{m s}^{-1}$$

The gun recoils at 0.4 m s⁻¹.

Note that the only forces acting are *internal* forces arising from the release of energy during the combustion of the explosive charge of the shell.

## A THEORETICAL PROOF OF THE PRINCIPLE OF CONSERVATION OF MOMENTUM

Fig. 3.7 shows two bodies, A and B, colliding. Body A exerts an 'internal' force $F'$ to the right on body B. Thus, according to Newton's third law, body

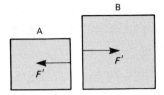

Fig. 3.7   Two bodies colliding

B will exert an equal force $F'$ to the left on body A. The time $t$ for which a force acts on each body will be the same — that is, the time of contact of the bodies.

From equation [3.3]

$$\text{(Change of momentum of body B)} = F' \times t$$

and

$$\text{(Change of momentum of body A)} = -F' \times t$$

The change of momentum of body A is negative since the force acting on it is to the left. Force, impulse and momentum are all vector quantities and we have assumed quantities acting to the right are positive. Hence:

$$\begin{pmatrix} \text{Total change in momentum} \\ \text{of objects A and B} \end{pmatrix} = \begin{pmatrix} \text{Change of} \\ \text{momentum of A} \end{pmatrix} + \begin{pmatrix} \text{Change of} \\ \text{momentum of B} \end{pmatrix}$$

$$= -F' \times t \qquad + F' \times t$$

$$= 0$$

Thus there is no change of momentum provided no *external* forces act. The principle of conservation of momentum is proved from Newton's third law.

**EXERCISE 3** _____

1)   A body of mass 4 kg has a momentum of $30 \, \text{kg m s}^{-1}$. Calculate its velocity.

2)   A body of mass 2 kg moves with a velocity of $8 \, \text{m s}^{-1}$ to the right. Another body of mass 3 kg moves along the same line. If the total momentum of the two bodies is $7 \, \text{kg m s}^{-1}$, calculate the velocity of the second body.

3)   A stationary squash ball of mass 0.025 kg is hit with a racket and leaves the racket head with a speed of $10 \, \text{m s}^{-1}$. If the time of contact with the racket is 0.05 s, calculate the average force exerted on the ball.

4)   The valve of a gas cylinder is opened and gas escapes at a rate of 0.5 kg each second. If the velocity of the escaping gas is $30\,\mathrm{m\,s^{-1}}$, calculate the force exerted on the cylinder.

5)   A machine gun fires bullets at a frequency of 300 per minute. If the bullets have mass 0.02 kg and a speed of $900\,\mathrm{m\,s^{-1}}$, calculate the average force exerted by the gun on the person holding it.

6)   A truck of mass 1000 kg moving at $3\,\mathrm{m\,s^{-1}}$ overtakes and collides with a truck of mass 2000 kg moving at $2\,\mathrm{m\,s^{-1}}$ in the same direction. The trucks become coupled together. Calculate their common velocity.

7)   Repeat Question 6 but assume that the trucks are moving along the same line in *opposite* directions.

8)   A pile driver of mass 300 kg moving at $15\,\mathrm{m\,s^{-1}}$ hits a stationary stake of mass 20 kg. Find the common velocity of stake and pile if they move off together.

9)   A bullet of mass 15 g and moving with a velocity of $400\,\mathrm{m\,s^{-1}}$ collides with a stationary target of mass 2.0 kg. If the bullet becomes embedded in the target find the velocity acquired by the target, assuming it is free to move.

10)   A space satellite has total mass 1000 kg. If a portion of mass 50 kg is ejected at a velocity of $30\,\mathrm{m\,s^{-1}}$, calculate the recoil velocity of the remaining portion of the satellite. Neglect the initial velocity of the satellite.

11)   A radioactive nucleus of mass 235 units travelling at $400\,\mathrm{km\,s^{-1}}$ disintegrates into a nucleus of mass 95 units and a nucleus of mass 140 units. If the nucleus of mass 95 units travels backwards at $200\,\mathrm{km\,s^{-1}}$, what is the velocity of the nucleus of mass 140 units?

12)   A rocket moving with velocity $100\,\mathrm{m\,s^{-1}}$ explodes into two equal parts. If one half of the rocket has its velocity reduced to zero by the explosion, what is the velocity of the other half?

# 4

# MECHANICAL ENERGY

## ENERGY

Energy exists in various forms. The two basic types of mechanical energy are:

(a) *kinetic energy* (KE), which is energy due to the motion of a body, and

(b) *potential energy* (PE), which is energy possessed by a body by virtue of its relative position. A book on a shelf and a stretched spring both possess potential energy.

## WORK

Work is done when energy is transferred from one system to another. When a man lifts a load there is a transfer of energy from the man to the raised load. The energy transfer is effected by a force acting over a distance and equals the work done. Work is defined as

$$\begin{pmatrix} \text{Work done} \\ \text{(Joules)} \end{pmatrix} = \begin{pmatrix} \text{Force} \\ \text{(Newtons)} \end{pmatrix} \times \begin{pmatrix} \text{Distance moved in direction of force} \\ \text{(metres)} \end{pmatrix}$$

### WORKED EXAMPLE 1

Refer to Fig. 4.1. A force of 300 N is applied over a distance of 4 m to the body. Calculate the energy transfer.

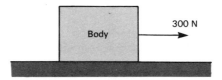

Fig. 4.1    Information for Worked Example 1

*Solution*

$$\text{Work done} = \text{Force} \times \text{Distance} = 300 \times 4$$
$$= 1200\,\text{J}$$
∴    $$\text{Energy transfer} = \text{Work done}$$
$$= 1200\,\text{J}$$

If there are no frictional forces and since the surface is horizontal, then all the energy transferred becomes moving or kinetic energy of the body.

**WORK DIAGRAMS**

Fig. 4.2 is a force versus distance graph for a constant force $F$ acting over a distance $d$. Now

(Energy transferred by force) = Work done = Force × Distance

$$= F \times d$$

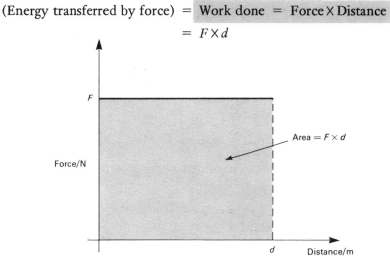

Fig. 4.2     A graph of force versus distance

But                    $F \times d$ = Area under graph of force versus distance

∴          Energy transfer = Work done = $\begin{pmatrix} \text{Area under graph of} \\ \text{force versus distance} \end{pmatrix}$

This is true irrespective of the shape of the graph.

*WORKED EXAMPLE 2*

Fig. 4.3 is a graph of force versus extension for a spring. Calculate the energy stored in the spring when it is extended by 30 cm.

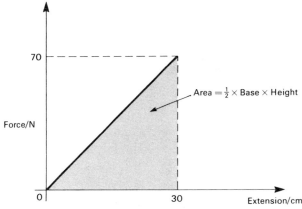

Fig. 4.3     Information for Worked Example 2

*Solution*

The work done by the force is stored as potential energy in the spring.

$$\text{Work done} = \text{Area under graph of force versus distance}$$

$$= \tfrac{1}{2} \times 70 \times 0.3 = 10.5$$

∴                 Energy stored = 10.5 J

## POTENTIAL ENERGY

The energy possessed by a body due to its position is called *potential energy* (PE in short).

Work is done in raising a body since we must apply a force, equal to the earth's gravitational attractive force, over some distance. The work done is stored as potential energy. We say a body raised above the Earth possesses *gravitational* potential energy because of its relative position in the Earth's gravitational field.

Similarly, since work is done in stretching, or compressing, a spring then a stretched, or compressed, spring posesses potential energy (see Worked Example 2). In this case the potential energy is the energy possessed due to the state of strain of the body.

### WORKED EXAMPLE 3

Calculate the gravitational potential energy possessed by a mass of 5 kg at a height of 2 m above the earth's surface.

*Solution*

A mass of 5 kg has a force of $5 \times 9.8$ equals 49 N acting on it due to the Earth's gravitational field (see Chapter 1). To raise it through 2 m, then

$$\text{Work done} = \text{Force} \times \text{Distance} = 49 \times 2$$

$$= 98 \text{ J}$$

The mass possesses gravitational potential energy of 98 J.

### GRAVITATIONAL POTENTIAL ENERGY

Fig. 4.4 shows a mass $m$ (kg) which has been raised a height $h$ (m) above a datum AA, such as the Earth's surface. From Chapter 1 the force required to raise the mass at constant speed  equals its weight $mg$, where $g$ is the gravitational field strength. Thus

$$\text{Work done in raising body a height } h = \text{Force} \times \text{Distance}$$

$$= mg \times h$$

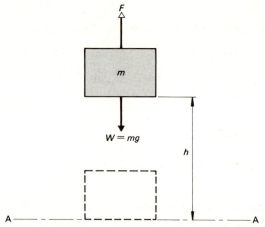

Fig. 4.4  Potential energy is energy possessed by a body by virtue of its position above a datum

An amount $mgh$ of energy is transferred from the lifting device to the body which gains gravitational potential energy due to its elevated position. Thus

$$\text{Gravitational PE} = mgh \qquad [4.$$

Gravitational PE is energy possessed by virtue of the position of the be above a chosen datum such as the Earth's surface, a laboratory bench Note that since height $h$ is measured from some arbitrary level gravitational PE is relative to this level.

## KINETIC ENERGY

*Kinetic energy* (KE in short) is the energy a body possesses due to its To make a stationary body move we apply a force over some dista energy transferred to the body which results in an increase in it called the kinetic energy of the body.

Fig. 4.5 shows a body of mass $m$ to which is applied a force $F$ over a distance $s$ and for a time $t$. Suppose the body is initially suppose it acquires a velocity $v$ as a result of the work done on body moves along a frictionless horizontal surface then all th becomes kinetic energy of the body. Now

$$\text{Distance travelled, } s = \text{Average velocity} \times \text{Time}$$

$$= \tfrac{1}{2}(0 + v) \times t$$

$$= \tfrac{1}{2}vt$$

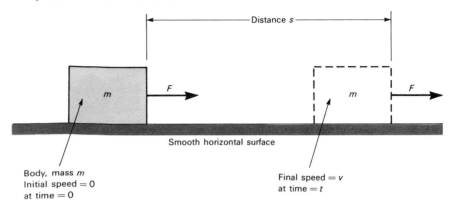

**Fig. 4.5**    Increasing the kinetic energy of a body

$$\text{Acceleration of body, } a = \frac{\text{Change in velocity}}{\text{Time}} = \frac{v-0}{t}$$

$$= \frac{v}{t}$$

or $$t = \frac{v}{a} \qquad [4.3]$$

Combining equations [4.2] and [4.3] gives

$$s = \tfrac{1}{2}vt = \tfrac{1}{2}v \times \frac{v}{a}$$

$$= \tfrac{1}{2}\frac{v^2}{a}$$

From equation [1.6] the acceleration $a$ is related to force $F$ and mass $m$ by

$$F = ma \qquad [1.6]$$

Thus                Work done on body $= F \times s$

$$= ma \times \tfrac{1}{2}\frac{v^2}{a}$$

$$= \tfrac{1}{2}mv^2$$

The work done becomes kinetic energy of the body. Therefore

$$\text{Kinetic energy} = \tfrac{1}{2}mv^2 \qquad [4.4]$$

*WORKED EXAMPLE 4*

A truck has a mass of 3 tonnes. If it is moving at $45 \, \text{km h}^{-1}$, how much kinetic energy does it possess?

*Solution*

We have

$$m = 3 \, \text{tonnes} = 3000 \, \text{kg}$$

$$v = 45 \, \text{km h}^{-1} = \frac{45 \times 1000}{60 \times 60} = 12.5 \, \text{m s}^{-1}$$

$$\text{Kinetic energy possessed by the truck} = \tfrac{1}{2}mv^2$$

$$= \tfrac{1}{2} \times 3000 \times 12.5^2$$

$$= 234\,000 \, \text{J} = 234 \, \text{kJ}$$

*WORKED EXAMPLE 5*

A body of mass $15 \, \text{kg}$ is on a horizontal frictionless surface. A force of $30 \, \text{N}$ acts on the body, initially at rest, and accelerates it to a final velocity of $12 \, \text{m s}^{-1}$. Calculate: (a) the acceleration, (b) the time taken to reach $12 \, \text{m s}^{-1}$, (c) the distance travelled in this time, (d) the work done by the force, (e) the final kinetic energy of the mass.

*Solution*

(a)  Acceleration $a = \dfrac{F}{m} = \dfrac{30}{15}$

$$= 2 \, \text{m s}^{-2}$$

(b)  An acceleration of $2 \, \text{m s}^{-2}$ means every second its speed increases by $2 \, \text{m s}^{-1}$. So it takes $6 \, \text{s}$ to reach $12 \, \text{m s}^{-1}$ from rest.

(c)  Distance travelled $s = \text{Average speed} \times \text{Time}$

$$= \tfrac{1}{2}(0 + 12) \times 6$$

$$= 36 \, \text{m}$$

(d)  Work done $= \text{Force} \times \text{Distance} = 30 \times 36$

$$= 1080 \, \text{J}$$

(e)  Kinetic energy $= \tfrac{1}{2}mv^2 = \tfrac{1}{2} \times 15 \times 12^2$

$$= 1080 \, \text{J}$$

The answers to (d) and (e) are equal since all the work done by the force becomes kinetic energy of the body. This is because the body moves along a horizontal frictionless surface. If the surface were inclined and/or rough, then some of the work done would become gravitational PE and/or energy dissipated to the surroundings as a result of work done against frictional forces.

## THE PRINCIPLE OF CONSERVATION OF ENERGY_____

This states that the total amount of energy in a given system is constant.

Although the total energy is always the same it may be converted from one form to another. The following are examples of energy conversion:

(a)    a body falling through a vacuum:

$$\text{Gravitational PE} \xrightarrow{\text{converted to}} \text{KE of moving body}$$

(b)    a body falling through a resisting medium such as a fluid:

$$\text{Gravitational PE} \xrightarrow{\text{converted to}} \text{KE + Energy dissipated via friction forces}$$

For a system such as (a) above in which no energy is dissipated via external forces then

$$\text{KE + PE} = \text{Constant}$$

This is called the *principle of conservation of mechanical energy.*

Fig. 4.6 shows a body of mass $m$ positioned at a height $h$ above a datum A–A. If it is allowed to fall it reaches its maximum velocity just as it arrives at the datum A–A. That is, the body attains its maximum kinetic energy when its potential energy is zero. Thus

$$\text{Loss in potential energy} = \text{Gain in kinetic energy} \qquad [4.5]$$

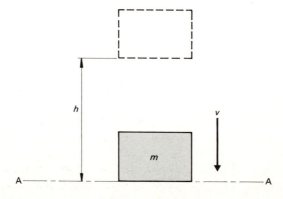

Fig. 4.6    A body falling from a height loses potential energy but gains kinetic energy

## WORKED EXAMPLE 6

A mass of 10 kg falls from a height of 45 m to ground level. Calculate: (a) the initial potential energy of the body, (b) the kinetic energy of the body as it hits the ground, (c) the velocity of the body as it hits the ground. (Assume $g = 9.8 \text{ m s}^{-2}$.)

*Solution*

(a)  From equation [4.1]

$$\text{Potential energy} = mgh = 10 \times 9.8 \times 45$$
$$= 4410 \text{ J}$$

(b)  We assume that energy dissipated via forces due to air resistance is negligible so that the sum of KE + PE remains constant. Thus, using the law of conservation of mechanical energy, equation [4.5], gives

$$\text{Kinetic energy} = \text{Loss of potential energy}$$
$$= 4410 \text{ J}$$

(c)  From equation [4.4]

$$\tfrac{1}{2}mv^2 = 4410$$

$$\therefore \qquad v = \sqrt{\frac{2 \times 4410}{m}} = \sqrt{\frac{2 \times 4410}{10}}$$

$$= 29.7 \text{ m s}^{-1}$$

The body hits the ground with speed of 29.7 m s$^{-1}$.

## WORKED EXAMPLE 7

A ball of mass 0.16 kg is propelled vertically upward with an initial velocity of 25 m s$^{-1}$. If the ball reaches a maximum vertical height of 20 m, what is the 'loss' in its energy? Where does this 'lost' energy go?

*Solution*

The initial kinetic energy of the ball as it leaves ground level is found using equation [4.4].

$$\text{KE} = \tfrac{1}{2}mv^2 = \tfrac{1}{2} \times 0.16 \times 25^2$$
$$= 50 \text{ J}$$

The gravitational potential energy of the ball when it reaches its maximum height is, from equation [4.1]

$$\text{PE} = mgh = 0.16 \times 9.8 \times 20$$
$$= 31.4 \text{ J}$$

Note that the initial KE and final PE of the ball are not the same. This is because some of the initial energy has been dissipated since work is done against air resistance. This energy transfer or 'loss' equals $50 - 31.4 = 18.6$ J.

*WORKED EXAMPLE 8*

A steel ball having a mass of 100 grams falls from a height of 1.8 m on to a plate and rebounds to a height of 1.25 m. Determine: (a) the potential energy of the ball before the fall; (b) the kinetic energy possessed by the ball as it hits the plate; (c) the velocity of the ball as it hits the plate; (d) the kinetic energy of the ball as it leaves the plate on the rebound; (e) the velocity of the ball on rebound.

*Solution*

(a)   From equation [4.1]

$$\text{Potential energy} = mgh = 0.1 \times 9.8 \times 1.8 = 1.76 \, \text{J}$$

(b)   Using the law of the conservation of energy (equation [4.5]) gives

$$\text{Kinetic energy} = \text{Loss of potential energy} = 1.76 \, \text{J}$$

(c)   From equation [4.4]

$$\tfrac{1}{2}mv^2 = 1.76$$

$$v = \sqrt{\frac{2 \times 1.76}{m}} = \sqrt{\frac{2 \times 1.76}{0.1}} = 5.93 \, \text{m s}^{-1}$$

(d)   Again using the law of conservation of energy, for this stage in which the ball rises after rebound gives

$$\text{Loss of kinetic energy} = \text{Gain in potential energy}$$
$$\text{Kinetic energy at rebound} = mgh$$
$$= 0.1 \times 9.8 \times 1.25$$
$$= 1.23 \, \text{J}$$

Note we have assumed that air resistance is negligible so that mechanical energy is conserved during rise and fall.

Note from (b) and (d) that there is a net energy transfer of $1.76 - 1.23 = 0.53$ J. This energy 'loss' becomes sound energy, heat energy etc. during the collision of the ball with the plate.

Worked Examples 6, 7 and 8 describe situations in which the sum of KE + PE is or is not conserved. Mechanical energy is conserved in the absence of friction or dissipative forces, for example:

(a)   when bodies fall or rise under gravity through a medium in which the resistive forces can be neglected, or when bodies slide along smooth surfaces;

(b)   the oscillation of a mass tethered to a spring in which friction forces may be neglected and air resistance assumed negligible. Fig. 4.7 shows two examples.

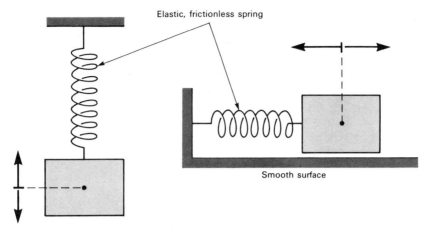

Fig. 4.7 Oscillations of a mass on an elastic spring

There is an interchange between kinetic energy (a maximum at the centre of oscillation) and elastic potential energy (a maximum at the edges of oscillation).

Mechanical energy is *not* conserved when friction or dissipative forces are significant. This is true when a body moves through a resistive medium or slides along a rough surface. When a snowflake falls through air the resistive forces are significant and the snowflake eventually reaches a constant, or terminal velocity. The oscillations of a mass on a spring can die away quickly if the mass is immersed in a liquid.

## ELASTIC AND INELASTIC COLLISIONS

In Chapter 3 we see that in a collision, provided that no external forces act, momentum is conserved. However, kinetic energy usually decreases during a collision. Since energy must be conserved we conclude that energy must be converted to other forms. Sound, heat and permanent deformation of the bodies are common examples.

An inelastic collision is one in which KE is *not* conserved.
An elastic collision is one in which KE *is* conserved.

Fig. 4.8 shows an inelastic collision in which 12 J of mechanical kinetic energy are converted to other forms. Some will become sound and heat energy. Most of it goes to deforming the plasticine. In a collision between two motor cars their bodywork could be deformed.

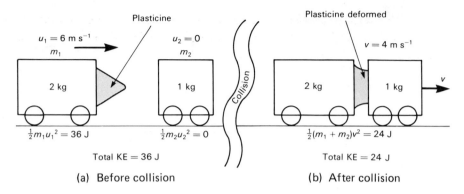

Fig. 4.8    An inelastic collision

An elastic collision is shown in Fig. 4.9. The spring is assumed to be elastic so that no energy is dissipated during its compression. At collision the spring is compressed. The potential energy is then converted to kinetic energy of the trolleys which acquire speeds as shown.

In practice perfectly elastic collisions between bodies of ordinary size are rarely achieved. In ideal gases the collisions of the molecules are perfectly elastic.

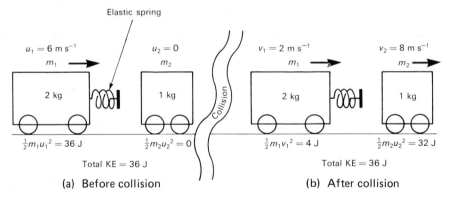

Fig. 4.9    An elastic collision

## EXERCISE 4

(Assume $g = 9.8 \, \mathrm{m\,s^{-2}}$.)

1)    A force of 200 N moves an object a distance of 25 m in the direction of the force and along a smooth horizontal surface. Calculate the energy gained by the object. What type of energy is this?

2)    Fig. 4.10 shows how the force on a planing machine tool varies during one cutting stroke. Calculate the energy transferred by the machine per stroke.

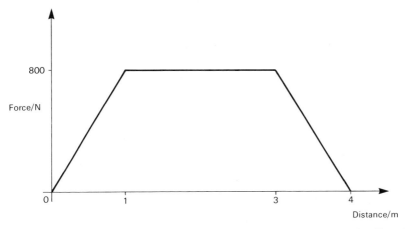

Fig. 4.10    Information for Question 2

3)    A vehicle has the following forces, in kilonewtons, exerted on it at the distances shown from its initial position:

| Distance/m | 0–20 | 20–60 | 60–100 | 100–150 |
|---|---|---|---|---|
| Force/kN | 6.0 | 5.0 | 4.0 | 3.5 |

Draw a work diagram and determine the work done by the force after: (a) 100 m, (b) 150 m.

4)    Calculate the gravitational potential energy possessed by an object of mass 200 kg at a height 12 m above ground level.

5)    Calculate the kinetic energy of: (a) a truck having a mass of 20 kg travelling at $3 \, m \, s^{-1}$; (b) a bullet with a mass of 2 g travelling at $400 \, m \, s^{-1}$; (c) a car with a mass of 1200 kg travelling at $70 \, km \, h^{-1}$.

6)    A mass of 25 kg is raised through a height of 8 m. What is the gain in gravitational potential energy?

7)    A body of mass 5 kg falls from rest and has a kinetic energy of 800 J just before touching the ground. (a) How much gravitational potential energy has been converted to kinetic energy? (b) From what height has the mass fallen?

8)    A truck having a mass of 10 tonnes is moving at $40 \, km \, h^{-1}$. How much kinetic energy does it possess? If it freewheels up an incline, what vertical height would it attain before coming to rest, assuming that there is no frictional resistance?

9)   A ball having a mass of 6 kg is dropped from a height of 25 m and rebounds with $\frac{3}{4}$ of its impact velocity. Find (a) the kinetic energy and (b) the velocity of the ball just before striking the ground, (c) determine the height to which the ball will rise after the first rebound.

10)   A body is projected with a velocity of 16 m s$^{-1}$ up the line of greatest slope of a smooth plane inclined at 30° to the horizontal. Calculate the distance travelled before the body comes to momentary rest.

11)   A grasshopper of mass 2.0 g has a vertical velocity of 3.0 m s$^{-1}$ when it jumps. Calculate its initial kinetic energy. What is the maximum vertical height it could reach. Why, in practice, is this not achieved?

12)   A ball of mass 0.20 kg drops from a height of 20 m and rebounds to a height of 7.2 m.

Calculate: (a) the energy dissipated on impact with the floor; (b) the velocity of the ball just before it hits the floor; (c) the velocity of the ball just after it hits the floor. (Assume the effects of air resistance may be neglected.)

13)   Fig. 4.11 shows an object of mass 5 kg being pulled up a smooth plane by a force of 15 N parallel to the plane. Find the kinetic energy of the object after sliding 20 m up the plane.

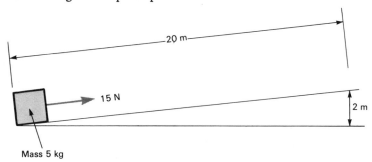

Fig. 4.11    Information for Question 13

14)   A body of mass 4 kg is projected with a velocity of 15 m s$^{-1}$ up the line of greatest slope of a *rough* plane inclined at 30° to the horizontal. If it travels 20 m before coming to momentary rest, calculate the energy dissipated via friction forces.

15)   Refer to Fig. 4.12. A truck of mass 100 kg is released from rest at A and moves along the frictionless track. Calculate its velocity at points B and C. Describe its subsequent motion and explain why it cannot travel beyond D.

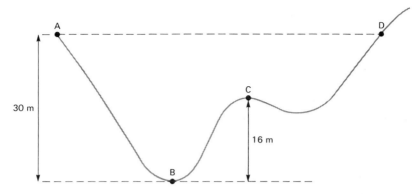

Fig. 4.12    Information for Question 15

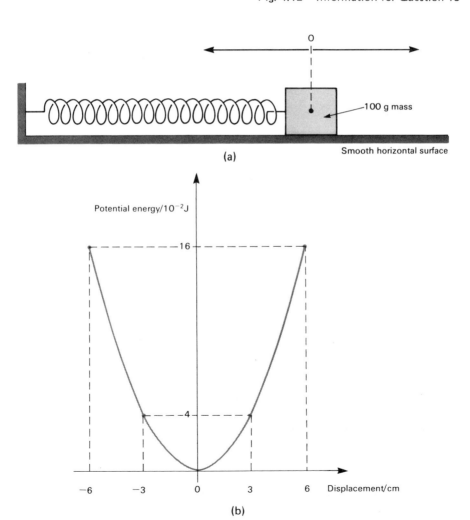

Fig. 4.13    Information for Question 16

**16)**  Fig. 4.13(a) shows a mass of 100 g tethered to the end of a horizontal and light elastic spring. The mass performs horizontal oscillations of maximum displacement 6 cm from the original position O. The graph of Fig. 4.13(b) shows how the elastic potential energy stored in the spring varies with displacement of the mass from 0. If the system has a total mechanical energy of 0.16 J; (a) explain why the total mechanical energy is constant; (b) calculate the kinetic energy and velocity of the mass when it passes through (i) 0 and (ii) a point 3 cm from 0.

**17)**  A 2 kg ball moving with a velocity of 8.0 m s$^{-1}$ collides with a 3 kg ball moving with a velocity of 6.0 m s$^{-1}$ moving in the same direction. If the balls stick together on impact, calculate the decrease in kinetic energy during the collision.

**18)**  Repeat Question 14 but with the balls moving in *opposite* directions.

**19)**  Fig. 4.14 shows two trolleys of mass 1.0 kg and 2.0 kg which are held together against a spring which is compressed. When the spring is released the trolleys separate and the 1.0 kg trolley moves to the left with a speed of 0.8 m s$^{-1}$. Calculate the initial total kinetic energy of the trolleys. Where does this energy come from?

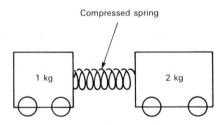

Compressed spring

1 kg          2 kg

Fig. 4.14    Information for Question 19

**20)**  A ball of mass 2.0 kg moving at 8.0 m s$^{-1}$ catches up with another ball of mass 3.0 kg moving at 6.0 m s$^{-1}$ in the same direction. If the collision is now *perfectly elastic*, calculate the velocity of *each* ball after the collision.

# 5 SOLID MATERIALS

## MATERIALS

The solid materials in common use can be grouped under three headings:

(a) the ferrous metals, which are composed mainly of iron and include all the steels, cast iron and wrought iron;

(b) the non-ferrous metals, which do not contain iron or contain it in very small quantities — for example, copper, brass, aluminium, magnesium, zinc, tin, titanium and the alloys of these metals;

(c) non-metallic materials, which include plastics, leather, rubber, glass, concrete, diamond and a variety of adhesives to name but a few.

## PROPERTIES OF MATERIALS

Each material has certain properties or qualities as follows:

(a) strength, the ability to resist force — iron, steel and some aluminium alloys;

(b) ductility, the ability to be drawn out into threads or wire — wrought iron, low carbon steels, copper, brass, aluminium;

(c) malleability, the ability to be rolled out into sheets or shaped by hammering — gold, copper, lead;

(d) brittleness, the tendency to break easily or suddenly with little or no permanent extension — cast iron, high-carbon steel, concrete;

(e) toughness, the ability to withstand suddenly applied or shock loads — certain alloy steels, some plastics, rubber;

(f) elasticity, the property which enables a body to return to its original shape when forces which have distorted it are removed — rubber, mild and medium carbon and alloy spring steels;

(g) hardness, the ability to resist surface indentation, wear or abrasion — high-carbon steels, some alloy steels, carborundum, diamond.

# DIRECT TENSILE AND COMPRESSIVE FORCES

Although there are many ways in which forces can be applied to a component the simplest forms of application are those which have a direct effect. We deal with two types, as follows:

(a)    Forces which pull and so tend to stretch the material are called *tensile forces*.

(b)    Forces which push and so tend to shorten the material are called *compressive forces*.

For a tensile or compressive force to have only a direct effect its line of action must pass through the centre of area of each cross-section of the component. That is, it must act axially otherwise some other effect such as bending will occur (see Fig. 5.1).

# STRESS AND STRAIN

If a tensile force is directly applied, it will distribute itself uniformly over each cross-sectional area. The amount of force transmitted per unit of area is called the tensile stress. Thus

$$\text{Tensile stress} = \frac{\text{Tensile force}}{\text{Cross-sectional area}} \qquad [5.1]$$

This stress causes the component to stretch in the direction of the force. The extension per unit length is called tensile strain. Thus

$$\text{Tensile strain} = \frac{\text{Extension}}{\text{Original length}} \qquad [5.2]$$

Let    $F$ = the applied direct tensile force (N)
$A$ = area of the cross-section ($m^2$)
$l$ = original length of the component (m)
$x$ = extension caused by $F$(m)
$\sigma$ = tensile stress ($N\,m^{-2}$)
$\epsilon$ = tensile strain

Then, from equation [5.1],

$$\sigma = \frac{F}{A} \qquad [5.3]$$

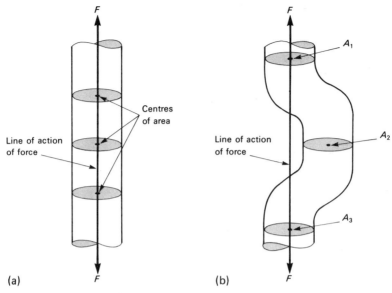

Fig. 5.1    Tensile forces: (a) direct loading, (b) indirect loading (bending would occur on sections such as $A_2$ where the line of action does not pass through the centre area)

and from equation [5.2],

$$\epsilon = \frac{x}{l} \qquad [5.4]$$

Since both $x$ and $l$ are measured in metres the units cancel and hence strain is simply a numerical quantity, that is, it has no units.

## COMPRESSIVE STRESS AND STRAIN

When a direct compressive force $F$ is applied to a component the latter will shorten slightly in the direction of the force. As in the case of tension, the compressive force will be distributed uniformly over the cross-sectional area withstanding the thrust, then

$$\text{Compressive stress } \sigma = \frac{F}{A} \qquad [5.5]$$

and

$$\text{Compressive strain } \epsilon = \frac{x}{l} \qquad [5.6]$$

where $x$ = reduction in length caused by the compressive force $F$.

### WORKED EXAMPLE 1

A cylindrical bar has a cross-sectional area of 400 square millimetres and a length of one metre. A direct tensile force of 16 000 N applied to the bar causes it to extend in length by 0.2 mm. Determine the stress in the bar and the strain.

*Solution*

$$\text{Tensile stress} = \frac{\text{Load}}{\text{Area of cross-section}}$$

or, using equation [5.3],

$$\sigma = \frac{F}{A}$$

$$= \frac{16\,000}{400}$$

$$= 40\,\text{N mm}^{-2} \quad \text{or} \quad 40\,\text{MN m}^{-2}$$

$$\text{Tensile strain} = \frac{\text{Extension}}{\text{Original length}}$$

or, using equation [5.4],

$$\epsilon = \frac{x}{l}$$

$$= \frac{0.2 \times 10^{-3}}{1}$$

$$= 0.000\,2$$

## TENSILE TESTING

The methods for the tensile testing of metals are specified in *British Standard 18*. Part 1 deals with non-ferrous metals and Part 2 deals with steel.

The cross-section of the test-piece may be circular, square, rectangular or annular. However the commonest type is the circular section test-piece (Fig. 5.2) which has the gauge length $l$, equal to 5 times its diameter.

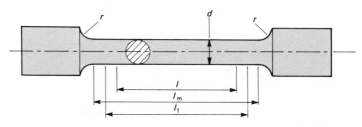

Fig. 5.2    The standard circular test-piece

Test-pieces vary in size and Table 5.1 shows the dimensions of British Standard test-pieces.

TABLE 5.1

| Cross-sectional area $A$ (mm$^2$) | Diameter $d$ (mm) | Gauge length $l$ (mm) | Minimum parallel length $l_{\mathrm{m}}$ (mm)* | Minimum transition radius $r$ (mm) | Tolerance on diameter ($\pm$ mm) |
|---|---|---|---|---|---|
| 400 | 22.56 | 113 | 124 | 23.5 | 0.13 |
| 200 | 15.96 | 80 | 88 | 15 | 0.08 |
| 150 | 13.82 | 69 | 76 | 13 | 0.07 |
| 100 | 11.28 | 56 | 62 | 10 | 0.06 |
| 50 | 7.98 | 40 | 44 | 8 | 0.04 |
| 25 | 5.64 | 28 | 31 | 5 | 0.03 |
| 12.5 | 3.99 | 20 | 22 | 4 | 0.02 |

*The gauge length is given to the nearest 1 mm and the minimum parallel length is adjusted accordingly.

**METHOD OF TESTING**

The test-piece is held in a testing machine by means of wedges, shouldered holders, etc. (see Fig. 5.7). It is pulled slowly in a controlled manner with the force applied axially to give direct loading. The test-piece is usually pulled until fracture occurs and the increasing values of load and the corresponding extensions are noted so that a load–extension graph can be drawn. A load–extension graph for wrought iron and low carbon steel is shown in Fig. 5.3.

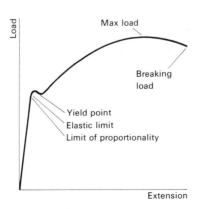

Fig. 5.3   A load–extension graph for wrought iron and low carbon steel

# ELASTIC AND PLASTIC DEFORMATION

At zero load there is, of course, zero extension. As the load is increased the extension at first increases proportionately so that when load and extension values are plotted a straight line is obtained. The relationship continues up to a point, called the *limit of proportionality*, beyond which the graph ceases to be a straight line. The *proportional limit* is that point up to which load is proportional to extension.

Just beyond this point is a second limit called the *elastic limit*.

The *elastic limit* is that point beyond which the specimen does not return to its original dimensions when the load is removed.

Up to this limit the specimen is completely elastic, so that if the load were removed the test-piece would return to its original length. Changes in dimensions which occur up to the elastic limit are called *elastic deformations*.

For loads which exceed the elastic limit the specimen becomes permanently, or plastically, deformed. Fig. 5.4 shows the intial part of the load–extension curve of a metal specimen. Up to the elastic limit L the specimen reverts to its original dimensions, so that when unloading the load extension data lie on the line LPO. However, if the specimen is loaded beyond L to a point such as S then during unloading the load extension data lie on the line SX. At X the specimen has a permanent extension $e_p$, sometimes called the *permanent set*. Changes in dimension which remain when the load is removed are termed *plastic deformations*.

If the specimen is again loaded, then the load extension data follow the line XS until at S further plastic deformation may occur.

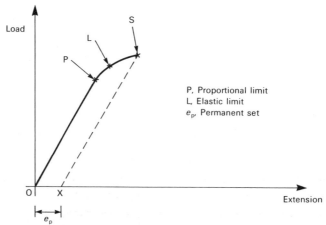

Fig. 5.4    Elastic and plastic deformation

When the elastic limit is being determined by experiment, great care must be taken to minimise any plastic deformation since this makes it difficult to determine L accurately.

## YIELD POINT

At a load a little higher than the elastic limit the test-piece may extend or yield suddenly by a comparatively large amount. This point is called the *yield point*. It shows as a distinct kink in the graph (see Fig. 5.3) for wrought iron and low-carbon steels but is not so evident in other metals.

## MAXIMUM LOAD

Further increases in load are now found to be necessary to cause further extension but equal increments of load cause greater amounts of extension and so the curve rises less and less steeply until the maximum load is reached (see Fig. 5.3).

Up to this point, the test-piece, as it is stretched, reduces only very slightly in diameter, an amount measurable only by micrometer or vernier, and so the cross-sectional area has remained virtually constant. Now, at the maximum load, a 'neck' or 'waist' begins to form locally, its exact location depending upon minute structural defects within the metal. From now on the cross-sectional area reduces rapidly (Fig. 5.5 overleaf) and so progressively less load is required to continue extending the test-piece.

## BREAKING LOAD

The curve of Fig. 5.3 is, therefore, now downwards as the extension continues, until eventually the test-piece breaks, usually with a cup and cone fracture as shown in Fig. 5.6. The load at which fracture occurs is called the *breaking load*.

# MEASURING EXTENSION

In most tensile test machines the extension is recorded by the machine, for example on a dial gauge or chart, is the sum of the deformations of the specimen *and of the machine itself* (for example, its weigh bar). To obtain accurate values of specimen extension we must eliminate the deformation of the machine. Up to, and just beyond the elastic limit the extension may be

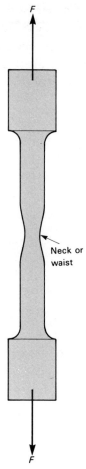

Fig. 5.5    A waist formed by extending beyond the maximum load

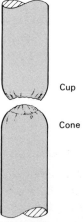

Fig. 5.6    A test-piece with a cup and cone fracture

measured using an extensometer (see Fig. 5.7). Since the instrument is clamped to the specimen, it allows the extension to be measured very accurately. It is then usual to remove this instrument to avoid damaging it when the specimen fractures. The larger extensions beyond the yield point are now measured by using a pair of dividers and a steel rule.

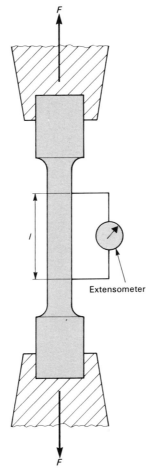

Fig. 5.7   Holding the test-piece in the testing machine

## STRESS–STRAIN GRAPHS

Following a tensile test the forces causing the extension of the test-piece can be divided by the original cross-sectional area and the resulting extensions by the original (gauge) length of the test-piece. Thus a set of tensile stresses and their corresponding strains are obtained and plotting these values gives a stress–strain diagram.

A stress–strain diagram is of more general use than a load–extension graph because the information it gives is not limited to the test-piece used but it can be applied to any component made of the same material as the test-piece. This is because stress refers to a standard area (1 m²) and strain is a fractional change in length. Thus stress and strain effectively refer to the same specimen size.

Stress–strain graphs are similar in shape to load–extension diagrams, and the stress–strain diagram for wrought iron and low-carbon steel will be similar in shape to Fig. 5.3. However, the important points on the graph give stresses instead of loads (Fig. 5.8). When the original cross-sectional area is used to calculate stress values these are termed nominal stress (see Fig. 5.8).

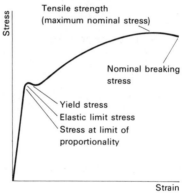

Fig. 5.8    A nominal stress–strain diagram for mild steel

Fig. 5.8 shows the breaking stress as being less than the maximum stress. This is impossible because the breaking stress must be the greatest stress to which the test-piece is subjected. The reason is that the cross-section does not remain constant as was assumed when calculating the stresses but it reduces in area as the test-piece is stretched. The reduction in area is not significant up to the yield point and it is very slight up to the point of maximum load. After this point the test-piece begins to form a waist and from then on the area reduces rapidly up to the point of fracture. If the loads had been divided by the *actual* cross-sectional areas to give *true* stresses, a graph like Fig. 5.9 would be obtained, and we see that the true breaking stress is the greatest stress attained during the test.

The stress corresponding to the maximum load is called the *tensile strength* and it is useful for comparing the strengths of various materials. In engineering design stress values beyond the tensile strength are never used and for all practical purposes the stress–strain curve of Fig. 5.8 is used instead of that of Fig. 5.9.

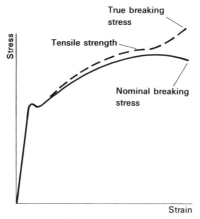

Fig. 5.9   A true stress–strain diagram for mild steel

Machinery and equipment are never operated beyond the yield stress of the material because permanent distortion would then occur. In fact, designers often work to 50% or less of the yield stress to take account of unusual loading conditions which may occur when the equipment is operating.

### WORKED EXAMPLE 2

A tensile test-piece has a cross-sectional area of 400 mm² and a gauge length of 113 mm. It is loaded until fracture occurs, and the loads and corresponding extensions are shown in Table 5.2.

(a)   Calculate the corresponding stress and strain values.

(b)   Plot the curve of stress versus strain, and from it determine the proportional limit and probable elastic limit.

TABLE 5.2

| Load/kN | Extension/mm | Load/kN | Extension/mm |
|---------|--------------|---------|--------------|
| 0 | 0 | 148 | 0.40 |
| 20 | 0.028 | 150 | 0.45 |
| 40 | 0.055 | 160 | 0.66 |
| 60 | 0.084 | 170 | 0.80 |
| 80 | 0.110 | 180 | 1.00 |
| 100 | 0.140 | 190 | 1.28 |
| 120 | 0.168 | 200 | 2.0 |
| 140 | 0.20 | 180 | 3.2 |
| 148 | 0.30 | 180 | Fracture |

TABLE 5.3

| Stress/$10^7$ N m$^{-2}$ | Strain/$10^{-3}$ | Stress/$10^7$ N m$^{-2}$ | Strain/$10^{-3}$ |
|--------------------------|------------------|--------------------------|------------------|
| 0 | 0 | 37.0 | 3.54 |
| 5.0 | 0.25 | 37.5 | 3.98 |
| 10.0 | 0.49 | 40.0 | 5.84 |
| 15.0 | 0.74 | 42.5 | 7.08 |
| 20.0 | 0.97 | 45.0 | 8.85 |
| 25.0 | 1.24 | 47.5 | 11.3 |
| 30.0 | 1.49 | 50.0 | 17.7 |
| 35.0 | 1.77 | 45.0 | 28.3 |
| 37.0 | 2.65 | 45.0 | Fracture |

### Solution

(a)   From equation [5.1], since the cross-sectional area is 400 mm² $= 4 \times 10^{-4}$ m² then

$$\text{Stress } \sigma = \frac{\text{Load}}{4 \times 10^{-4}} \text{N m}^{-2}$$

From equation [5.2], since the original length is 113 mm then

$$\text{Strain } \epsilon = \frac{\text{Extension (mm)}}{113}$$

Values of stress and strain are shown in Table 5.3.

(b)    The stress–strain curve is shown in Fig. 5.10. It ceases to be a straight line at about $35 \times 10^7 \, \text{N m}^{-2}$, so that this is the limit of proportionality. This also gives an approximate value for the elastic limit stress. Careful experimental work may lead to a more precise answer.

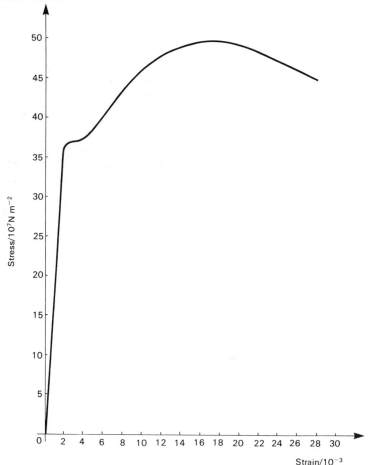

Fig. 5.10    Stress–strain curve for specimen of Worked Example 2

## THE MODULUS OF ELASTICITY

Most engineering materials display some elasticity when loaded. Up to the proportional limit of the elastic region the load is found to be proportional to the extension so that when these two quantities are plotted against each other a straight line is produced. Similarly stress is proportional to strain up to the limit of proportionality (which is approximately the elastic limit stress) for the material. Thus

$$\text{Stress} \propto \text{Strain}$$

or

$$\text{Stress} = \text{A constant} \times \text{Strain}$$

or

$$\frac{\text{Stress}}{\text{Strain}} = \text{A constant}$$

The constant is known as the *modulus of elasticity* or the *elastic modulus*. If the specimen is in tension or compression, it is called *Young's modulus* and is denoted by the letter $E$, so that

$$\frac{\text{Stress}}{\text{Strain}} = E \qquad [5.7]$$

It follows from equation [5.7] that $E$ is represented by the slope of the straight line portion of the stress–strain graph since the slope is given by the ratio of the vertical (stress) ordinate to the corresponding horizontal (strain) ordinate.

The unit of stress is $N\,m^{-2}$, and as strain has no units then the unit of $E$ must be $N\,m^{-2}$. Different materials have different values of $E$, so that the modulus of elasticity when found can be useful in determining the kind of material from which a component is made. Values of $E$ tend to be high and are usually expressed in $GN\,m^{-2}$ (giganewtons per metre squared, where the prefix giga, or G, represents the factor $10^9$). Thus $E$ for steel is about $207 \times 10^9\,N\,m^{-2}$ or $207\,GN\,m^{-2}$. Table 5.4 lists approximate values of $E$ for various materials.

TABLE 5.4

| Material | $E\,(GN\,m^{-2})$ |
|---|---|
| Wrought iron | 186 |
| Steel | 207 |
| Cast iron | 96 |
| Aluminium and its alloys | 70 |
| Copper | 124 |
| Brass | 103 |
| Phosphor bronze | 96 |
| Magnesium and its alloys | 45 |
| Beryllium | 300 |
| Titanium alloys | 100 |
| Monel | 175 |
| Carbon fibre | 130–190 |
| Nylon | 2.75 |
| Rubber | 0.035 |
| Polystyrene | 2.4–4.1 |
| Polythene (low-density) | 0.117–0.240 |
| Polythene (high-density) | 0.550–1.030 |

## WORKED EXAMPLE 3

Calculate Young's modulus for the specimen of Worked Example 2.

*Solution*

Young's modulus is the slope of the straight line portion of the stress–strain graph. The graph is a straight line up to a stress of $30 \times 10^7 \, \text{N m}^{-2}$ and a strain of $1.49 \times 10^{-3}$.
From equation [5.7]

$$\text{Young's modulus, } E = \frac{\text{Stress}}{\text{Strain}} = \frac{30 \times 10^7}{1.49 \times 10^{-3}}$$

$$= 201 \, \text{GN m}^{-2}$$

## WORKED EXAMPLE 4

A steel bar is of length 0.50 m and has a rectangular cross-section 15 mm by 30 mm. Calculate the force required to extend it by 0.2 mm given that Young's modulus for steel is 200 GN m$^{-2}$. Assume that the proportional limit is not exceeded.

*Solution*

Suppose a force $F$ is required. From equations [5.7], [5.1] and [5.2]

$$E = \frac{\text{Stress}}{\text{Strain}} = \frac{(\text{Force} \div \text{Area})}{(\text{Extension} \div \text{Original length})}$$

We have

$$E = 200 \times 10^9 \, \text{N m}^{-2}$$

$$\text{Area} = 15 \times 10 = 150 \, \text{mm}^2 = 150 \times 10^{-6} \, \text{m}^2$$

$$\text{Extension} = 0.2 \, \text{mm} = 0.2 \times 10^{-3} \, \text{m}$$

and          Original length = 0.5 m

Therefore

$$200 \times 10^9 = \frac{F \div (150 \times 10^{-6})}{(0.2 \times 10^{-3}) \div (0.5)}$$

Rearranging gives

$$F = 36 \times 10^3 \, \text{N} = 36 \, \text{kN}$$

## WORKED EXAMPLE 5

An aluminium alloy strut in the landing gear of an aircraft has a cross-sectional area of 40 mm$^2$ and a length of 750 mm. As the aircraft lands the strut is subjected to a compressive force of 2800 newtons. Calculate by how much the strut will shorten under this force. Assume that the proportional limit is not exceeded.

*Solution*

$$\text{Compressive stress } \sigma = \frac{\text{Load}}{\text{Area}}$$

$$= \frac{2800}{40}$$

$$= 70 \, N \, mm^{-2}$$

$$= 70 \, MN \, m^{-2}$$

From equation [5.7]

$$\text{Strain} = \frac{\text{Stress}}{E}$$

From Table 5.3, $E$ for aluminium alloy is 70 GN m$^{-2}$ or $70 \times 10^9$ N m$^{-2}$, then

$$\text{Strain } \epsilon = \frac{70 \times 10^6}{70 \times 10^9}$$

$$= 0.001$$

Also from equation [5.6]

$$\epsilon = \frac{\text{Reduction in length}}{\text{Original length}}$$

Therefore

$$\text{Reduction in length} = 0.001 \times 750$$

$$= 0.750 \, mm$$

# MORE STRESS–STRAIN CHARACTERISTICS

So far we have discussed stress–strain diagrams for wrought iron and the low-carbon steels. Other materials behave quite differently when undergoing a tensile test and the way in which some of the more common materials behave is discussed below.

## THE HIGHER-CARBON STEELS

As the percentage carbon in the steel increases the yield point becomes less definite, and hence more difficult to determine until the carbon content is around 1% when the metal no longer yields suddenly and the load–extension curve shows a smooth transition from the elastic to the plastic regions. The test-piece does not form a waist and fracture is sudden, occurring at the maximum load (Fig. 5.11).

## CAST IRON

Ordinary grey cast iron contains around 3.5% carbon, that is, more carbon than is contained in the steels where the upper limit is about 1.8%. Cast iron is very brittle and weaker in tension than the steels. It is not at all ductile

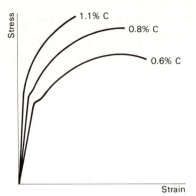

Fig. 5.11    A stress–strain graph for higher-carbon steels

and so does not form a waist before fracture. The stress–strain curve for cast iron is similar to that shown in Fig. 5.12.

Although cast iron is brittle, it has elastic properties. Extension is proportional to load within the limited elastic range indicated by the straight line portion of the graph. There is no yield point and fracture is sudden.

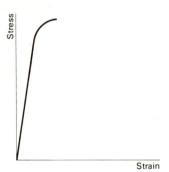

Fig. 5.12    A stress–strain graph for cast iron

## GLASS

Glass is a brittle material which fractures due to the propagation of minute cracks which exist in it. It exhibits negligible plastic deformation. It shows small elastic strains — typically stress is proportional to strain up to stresses around $10^8 \, \text{N m}^{-2}$ in tension, at which point it fractures. The form of the stress–strain curve is similar to that for cast iron (see Fig. 5.12).

## COPPER

When copper is annealed (softened by heating and subsequent cooling) it is very ductile and yet does not have a yield point. A test-piece develops a waist, reducing in cross-sectional area before fracture so that there is a smooth transition from the elastic to the plastic state. The curve reaches a maximum load then falls to a lower value before fracture takes place (Fig. 5.13).

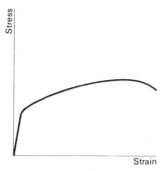

Fig. 5.13   A stress–strain graph for copper

## ALUMINIUM ALLOYS

The alloys of aluminium are numerous and varied. The stress–strain curve shown in Fig. 5.14 is probably typical of many normal structural aluminium alloys as rolled, extruded or drawn, such as duralumin whose average composition is aluminium 95%, copper 4%, magnesium 0.5% and manganese 0.5%. The extension is less than that of mild steel and there is no definite yield point. There is slight local reduction in cross-sectional area before fracture due to the formation of a waist.

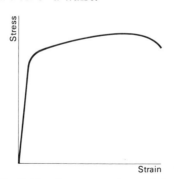

Fig. 5.14   A stress–strain graph for wrought aluminium alloys

## PLASTICS

Stress–strain curves for plastics are not so predictable as they are for metals. The rate at which the load is applied, the humidity of the atmosphere and sensitivity to relatively small changes in temperature can all affect the behaviour of a plastic. However, curves for plastics other than very hard brittle ones tend to approximate to the shape shown in Fig. 5.15 when the load is applied over a short period of time. This generalised stress–strain curve can vary as shown below depending upon the type of plastic.

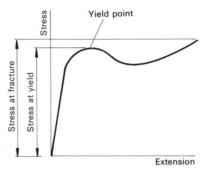

Fig. 5.15    A general stress–strain graph for plastics

### BRITTLE AND DUCTILE MATERIALS

Materials which exhibit a significant amount of plastic deformation are called *ductile materials*. A ductile material will have a stress–strain curve similar to that of mild steel (Fig. 5.8) or copper (Fig. 5.13) and aluminium (Fig. 5.14). Most common pure metals and alloys are ductile materials.

Ductile materials can plastically deform because their structure is such that the planes of atoms can relatively easily slide over each other. This occurs by the movement of whole lines of atoms, called dislocations.

Materials which exhibit little or no plastic deformation are called *brittle materials*. A brittle material will have a stress–strain curve similar to that of cast iron (Fig. 5.12). Glasses, ceramics and certain hard plastics are brittle materials. They fracture by splitting open due to cracks running through them. Their structure is such that dislocation motion is not possible, so the structure cracks open instead.

### EXERCISE 5

1)    A steel wire 6 metres long and 1.83 mm diameter is found to extend by 3.73 mm when a tensile load of 300 N is applied. Calculate: (a) the tensile stress; (b) the strain.

2)    Calculate the minimum diameter of round bar which is required to sustain a direct tensile load of 150 kN if the stress is not to exceed 124 MN m$^{-2}$.

**3)**   A test on a specimen of mild steel having a rectangular cross-section 75 mm by 25 mm gave an extension of 0.081 mm on a gauge length of 200 mm due to a load of 150 kN. Calculate the stress and the strain in the specimen.

**4)**   A hollow cast iron column of square cross-section has internal sides measuring 150 mm and it is to carry a load of 1000 kN. If the safe working stress is limited to 100 MN m$^{-2}$, calculate: (a) the area of metal required to carry the load; (b) the thickness of the walls of the column.

**5)**   A load of 120 N hangs from the end of a copper wire and produces an extension of 0.45 mm. If the diameter of the wire is 2.0 mm and its length is 1.5 m, calculate: (a) the tensile stress; (b) the tensile strain; (c) the value of Young's modulus for copper; (d) the maximum load the wire will support if copper has a tensile strength of $2.0 \times 10^8$ N m$^{-2}$.

Assume for part (c) that the proportional limit has not been exceeded.

**6)**   A metal wire 1.60 mm diameter and 1.83 m long is subject to a load of 44 N and the extension is found to be 1.14 mm. Find Young's modulus of elasticity.

**7)**   A steel bar of circular cross-section is 2.00 m long and carries a load of 120 kN. If the maximum permissible stress is not to exceed 90 MN m$^{-2}$, determine the minimum diameter of the bar. If Young's modulus of elasticity is 207 GN m$^{-2}$, calculate the extension of the bar.

**8)**   A hollow vertical cast iron column 1.80 m long has an outside diameter of 300 mm and an internal diameter of 250 mm. It carries a compressive load of 500 kN. If $E$ for cast iron is 96 GN m$^{-2}$, calculate: (a) the stress in the column; (b) the amount which the column will shorten.

**9)**   A mild steel bar has a rectangular cross-section of 90 mm by 20 mm. It is 12 m long and its maximum allowable extension is to be limited to 2.5 mm. If the modulus of elasticity of the metal is 207 GN m$^{-2}$, find the maximum load it will be able to carry.

**10)**   A short cast iron column is to be a hollow cylinder 150 mm external diameter and it is to support an axial load of 360 kN. If the permissible stress is not to exceed 90 MN m$^{-2}$, find the inside diameter.

**11)** A certain member of the structure of a bridge shortens by 0.25 mm when under load. The cross-sectional area of this member is 20 000 mm² and it is 2.00 metres long. Calculate: (a) the stress in the member; (b) the force in the member. Take $E$ for material to be 200 GN m$^{-2}$.

**12)** A vertical bar 3 m long carries a load of 20 kN. If the extension in the bar is not to exceed 0.20 mm, calculate: (a) the stress in the bar; (b) the minimum cross-sectional area of the bar. Take $E = 200$ GN m$^{-2}$.

**13)** A mild steel bar 15.96 mm in diameter and 80 mm long extended 0.060 mm under a tensile load of 30 kN. Calculate the value of $E$.

**14)** The following tensile test data were obtained using a mild steel specimen:

| Load/N | 0 | 100 | 200 | 300 | 400 | 500 | 600 |
|---|---|---|---|---|---|---|---|
| Extension/mm | 0 | 0.025 | 0.050 | 0.075 | 0.100 | 0.125 | 0.150 |

| Load/N | 700 | 800 | 900 | 1000 | 1050 | 1100 |
|---|---|---|---|---|---|---|
| Extension/mm | 0.175 | 0.200 | 0.230 | 0.280 | 0.460 | 0.660 |

Original length of specimen = 0.20 m
Original area of specimen = 4.0 mm² = $4.0 \times 10^{-6}$ m²

(a) Plot the load extension graph.
(b) Calculate the stress and strain values for each load.
(c) Plot the stress–strain graph and identify the proportional limit.
(d) If, on removal of the load of 11 N the stress–strain graph follows a line parallel to the original proportional portion, estimate the 'permanent' strain left in the specimen on removal of the load.

**15)** During a tensile test on a bronze specimen with a diameter of 15.96 mm and a gauge length of 80 mm the following results were obtained:

| Load/kN | 10 | 20 | 30 | 35 | 40 | 45 | 50 | 55 |
|---|---|---|---|---|---|---|---|---|
| Extension/mm | 0.04 | 0.08 | 0.12 | 0.14 | 0.164 | 0.20 | 0.35 | 0.62 |

Draw the load–extension graph using the following scales:
Load (vertical) 4 cm = 10 kN,
Extension (horizontal) 1 cm = 0.05 mm.
From the graph and by subsequent calculation obtain the value of $E$.

**16)** The results below were obtained during a tensile test using a test-piece with a diameter of 15.96 mm and a gauge length of 80 mm. Find the value of $E$.

| Load/kN | 5 | 10 | 15 | 17.5 | 20 | 22.5 | 25 | 27.5 | 30 |
|---|---|---|---|---|---|---|---|---|---|
| Extension/mm | 0.01 | 0.02 | 0.03 | 0.035 | 0.041 | 0.05 | 0.088 | 0.15 | 0.23 |

**17)** Sketch a stress–strain graph for glass given that glass obeys Hooke's law up to a stress of $0.5 \times 10^8 \, \text{N m}^{-2}$, at which point it fractures. Young's modulus for glass is $7.0 \times 10^{10} \, \text{N m}^{-2}$.

**18)** Buildings are often made of materials such as brick or stone or concrete which are all brittle. Why do you think this is so?

# 6

# IDEAL GASES

## GASES

In the simplest division between material forms we divide substances into *solids, liquids* and *gases*. This is too simple a division in the case of gas-like materials because of what are called *vapours*. For instance, at room temperature carbon dioxide should be classed as a vapour, not as a gas. The difference between a gas and a vapour, at a given temperature, is that the vapour can be liquefied by the application of pressure alone, whereas the gas cannot. Even if we ignore the problem of vapours the situation with regard to gases is still complicated, because unlike solids and liquids the volume of gas is strongly affected by the pressure a gas is subject to.

For a gas the possible physical quantities which govern its volume $V(m^3)$ are as follows:

(a)  The pressure $P$ it is subject to (unit $N\,m^{-2}$).

(b)  The temperature of the gas.

In the case of temperature we would normally measure this in degrees Celsius or Centigrade ($\theta$) but there is an excellent reason for using a different scale of temperature, degrees absolute (or Kelvin, K). Before looking at the connection between these two scales of temperature let us look at the behaviour of a gas in general.

Since there are *three* variables (pressure, volume and temperature) we need to hold one constant and look at the relationship between the other two.

## BOYLE'S LAW

In this case the *temperature* of the gas is kept constant and the relationship between pressure ($P$) and volume ($V$) is investigated. The resulting variation is shown in Fig. 6.1. We expect the volume to decrease as we increase the pressure on a gas. When the variation is presented in a different form the situation is much easier to understand (Fig. 6.2).

74

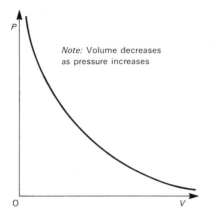

Fig. 6.1   Pressure versus volume

Pressure exerted is directly proportional to the inverse of the volume occupied:

$$P \propto \frac{1}{V}$$

Another method of stating the relationship is to imagine pressure as $y$ and $\frac{1}{\text{Volume}}$ as $x$. Since the result is a straight line through the origin, then

$$y = kx$$

where $k$ is a constant

or

$$P = k\left(\frac{1}{V}\right)$$

or

$$PV = k \qquad\qquad [6.1]$$

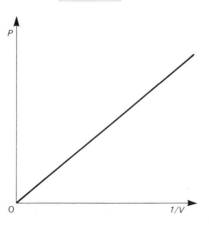

Fig. 6.2   Pressure versus 1/volume

For a fixed mass of gas at constant temperature the product of the volume a gas occupies, and the pressure it exerts is a constant. So if you keep the mass of gas constant, and you keep the temperature of the gas constant, then $PV$ = constant.

Thus if the pressure changes from $P_1$ to $P_2$, then the volume changes from $V_1$ to $V_2$ and

$$P_1V_1 = P_2V_2 \qquad\qquad [6.2]$$

### WORKED EXAMPLE 1

If at some fixed temperature the volume of a certain mass of gas is $2\,m^3$ and the pressure it exerts is $8\times10^3\,N\,m^{-2}$, what volume would it occupy if its pressure increased to $1\times10^5\,N\,m^{-2}$?

*Solution*

From equation [6.2]         $P_1V_1 = P_2V_2$

The initial state is          $P_1 = 8\times10^3\,N\,m^{-2}, \quad V_1 = 2\,m^3$

The final state is            $P_2 = 1\times10^5\,N\,m^{-2}, \quad V_2 = ?$

Hence                        $8\times10^3\times2 = 1\times10^5\times V_2$

or                $V_2 = \dfrac{8\times10^3\times2}{1\times10^5} = 0.16\,m^3$

The calculation is straightforward. The only question that might arise is why such large values for pressure? The answer is that normal atmospheric pressure varies around about $1\times10^5\,N\,m^{-2}$, and that in gas cylinders the pressure can be 100 times this value.

## BOYLE'S LAW STATEMENT

The variation of volume with pressure for a gas was first investigated by a scientist called Robert Boyle and so the modern statement of this behaviour bears his name. Boyle's law states:

If the mass and temperature of a gas are kept constant then the pressure a gas exerts is inversely proportional to the volume it occupies.

## CHARLES' LAW

Just as temperature could be held constant so the pressure a gas exerts can be held constant. The variables in this case are volume $V$ and temperature. At the moment that means $\theta$ is measured in degrees Celsius. The variation is shown in Fig. 6.3. The variation is seen to be linear but not so simple as the variation for Boyle's law in the sense that $V_0$ depends on the

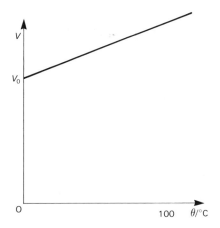

Fig. 6.3    Volume versus temperature ($V_0$ = volume at $0^\circ$C)

pressure. If the line is continued backward, however, it meets the temperature axis (see Fig. 6.4). This point of interception is independent of the fixed pressure and also of the mass of gas or in fact any other possibility. Note that at this temperature the gas would (theoretically) occupy zero volume. This point is called the *absolute zero of temperature* and is the lowest possible value of temperature.

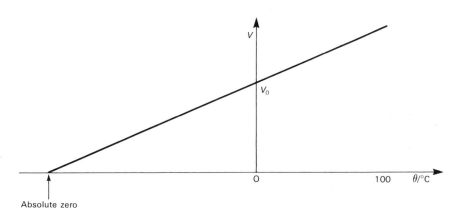

Fig. 6.4    Zero volume and absolute zero of temperature

The obvious next step is that if this is the lowest possible temperature then why not call it zero? This is what the Kelvin or absolute scale of temperature does. The symbol for temperature is now $T$ and it is measured in degrees Kelvin or degrees absolute (K). The *interval* on this scale is the same as the *interval* on the Celsius scale but the two scales have a different origin.

The effect of this is seen in Fig. 6.5. If we use absolute zero as our origin, then

$$V \propto T$$

or
$$\frac{V}{T} = \text{Constant} \qquad [6.3]$$

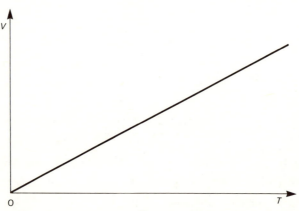

Fig. 6.5    Volume versus absolute temperature

Thus if the volume changes from $V_1$ to $V_2$ when the temperature changes from $T_1$ to $T_2$, then

$$\frac{V_1}{T_1} = \frac{V_2}{T_2} \qquad [6.4]$$

Note that we need to know the relationship between $T$ and $\theta$. This relationship is

$$T = \theta + 273$$

This means that the absolute zero of temperature occurs at $-273\,^{\circ}\text{C}$.

*Note.* If there were different intercepts for

(a)    different masses of the same gas,

(b)    different pressures of the same gas,

(c)    different values for different gases,

then the value of absolute zero would be of no significance. In fact we find the value of the intercept is *always* the same.

### WORKED EXAMPLE 2

At a certain pressure the volume of a gas is 4 m³ at 57°C. What volume would it occupy at 20°C?

*Solution*

We must first convert temperatures to degrees absolute, as follows:

$$57°C \equiv 273 + 57 = 330 \text{ K}$$
$$20°C \equiv 273 + 20 = 293 \text{ K}$$

Note we do not use °K but write K.

From equation [6.4]

$$\frac{V_1}{T_1} = \frac{V_2}{T_2}$$

The initial state is $V_1 = 4 \text{ m}^3$, $T_1 = 330 \text{ K}$
The final state is $V_2 = ?$, $T_2 = 293 \text{ K}$

Hence

$$\frac{4}{330} = \frac{V_2}{293}$$

or

$$V_2 = \frac{4 \times 293}{330} = 3.55$$

The final volume at 20 °C is 3.55 m³.

## STATEMENT OF CHARLES' LAW

The volume of a fixed mass of gas at constant pressure varies directly with the absolute temperature.

*Note.* If we superimpose Figs. 6.4 and 6.5 we get Fig. 6.6. This shows that the slope of the line is $\dfrac{V_0}{273}$ and is also equal to $\dfrac{V}{T}$, the constant in equation [6.3].

Thus

$$\frac{V_0}{273} = \frac{V}{T}$$

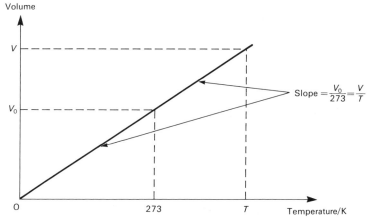

Fig. 6.6   Slope of volume versus temperature

or
$$V = \frac{V_0}{273}(T)$$

$$= \frac{V_0}{273}(273 + \theta)$$

or
$$V = V_0\left(1 + \frac{1}{273}\theta\right) \qquad [6.5]$$

Equation [6.5] tells us that for every degree rise in temperature the volume increases by $\dfrac{V_0}{273}$. This yields the more normal statement of Charles' law:

The volume of a fixed mass of gas at constant pressure increases by $\dfrac{1}{273}$ of its volume at 0 °C for every degree Celsius rise in temperature.

## THE PRESSURE LAW

The final variation of this section is when the volume is kept constant and the pressure and temperature are allowed to vary. The situation is virtually a repetition of Charles' law. The effect of varying the temperature and holding the volume constant is seen in Fig. 6.7. If the line is extended backwards, it again cuts the axis at $-273$ °C. The result therefore of a change of the temperature axis is as shown in Fig. 6.8. We see

$$P = \text{Constant} \times T$$

or
$$\frac{P}{T} = \text{Constant} \qquad [6.6]$$

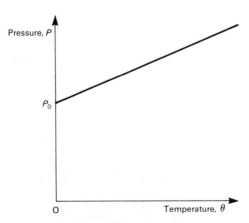

Fig. 6.7    Pressure versus temperature (degrees Celsius)

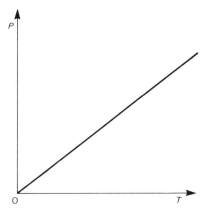

Fig. 6.8    Pressure versus absolute temperature

## STATEMENT OF THE PRESSURE LAW

The pressure of a fixed mass of gas at constant volume increases directly with the absolute temperature.

Alternatively, in a similar way to the above section, this can be restated as:

The pressure of a fixed mass of gas at constant volume increase by 1/273 of its pressure at $0\,^{\circ}C$ for every degree Celsius rise in temperature.

## IDEAL GAS

Rather than remember three equations for gas behaviour

$$PV = \text{Constant} \quad (\text{equation } [6.1])$$

$$\frac{V}{T} = \text{Constant} \quad (\text{equation } [6.3])$$

$$\frac{P}{T} = \text{Constant} \quad (\text{equation } [6.6])$$

we collect these all together into one equation and just remember

$$\frac{PV}{T} = \text{Constant} \qquad [6.7]$$

An alternative way of writing this is

$$\frac{P_1 V_1}{T_1} = \frac{P_2 V_2}{T_2}$$

[6.8]

where $P_1$, $V_1$, $T_1$ refer to the initial state and $P_2$, $V_2$, $T_2$ to the final state.

*Note*. If $P$ or $V$ or $T$ alone were constant, as in the three previous cases, then the fixed parameter would automatically cancel from the relationship of equation [6.7]. This would leave us with one of equations [6.1], [6.3] or [6.6].

### WORKED EXAMPLE 3

What is the mass of oxygen present in a cylinder of volume 5 litres at a temperature of 27°C and a pressure of 3 MN m$^{-2}$? Oxygen has a density of 1.43 kg m$^{-3}$ at **standard temperature and pressure** (0°C and $1.013 \times 10^5$ N m$^{-2}$).

*Solution*

Since we are given the density at STP we must find the volume at STP. We have

*Initial state*:    $P_1 = 3 \times 10^6$ N m$^{-2}$

$V_1 = 5$ litres $= 5000$ cm$^3 = 5 \times 10^{-3}$ m$^3$

$T_1 = 27°C = 300$ K

*Final state*:    $P_2 = 1.013 \times 10^5$ N m$^{-2}$

$V_2 = ?$

$T_2 = 0°C = 273$ K

Rearranging equation [6.8], we have

$$V_2 = \frac{P_1 V_1}{T_1} \times \frac{T_2}{P_2} = \frac{3 \times 10^6 \times 5 \times 10^{-3} \times 273}{300 \times 1.013 \times 10^5}$$

$$= 0.135$$

The volume at STP is 0.135 m$^3$. But

$$\text{Mass of gas} = \text{Volume} \times \text{Density}$$

$$= 0.135 \times 1.43$$

$$= 0.193$$

The cylinder contains 0.193 kg of oxygen.

The relationship of equations [6.7] and [6.8] works for all gases reasonably well, i.e. at low pressures and temperatures above the liquefaction point. It would work exactly for an ideal gas. An ideal gas is one that obeys the relationship perfectly. The statement involves a constant mass of gas. If mass is brought into the relationship of equation [6.7] it becomes

$$\frac{PV}{Tm} = \text{Constant}$$

where $m$ is the actual mass of gas contained in volume $V$. A better statement is

$$PV = \frac{m}{M} RT \qquad [6.9]$$

where $M$ is the molecular weight of the gas expressed in kilograms (for example, $M$ is 0.032 kg for oxygen). Note that $\frac{m}{M}$ is the number of moles of the gas trapped in the volume $V$. We usually denote the number of moles of the gas by letter $n$.

Thus equation [6.9] becomes

$$PV = nRT \qquad [6.10]$$

$R$ is the same for all gases and so is called the *universal gas constant*. It has the value 8.3 J mol$^{-1}$K$^{-1}$. Its units will now be explained. From equation [6.10]

$$R = \frac{PV}{nT}$$

where the units are: $P$, N m$^{-2}$; $V$, m$^3$; $n$, mole; and $T$, K. Thus $PV$ has units of newton-metre (N m) which, from Chapter 4, is equivalent to Joules (J). So the units of $R$ are J mol$^{-1}$K$^{-1}$.

## WORKED EXAMPLE 4

How many moles of gas are there present in a gas cylinder of volume 0.02 m$^3$ at a temperature of 20°C and a pressure of 15 MN m$^{-2}$? (Assume $R = 8.3$ J mol$^{-1}$ K$^{-1}$.)

*Solution*

First, place the information in the correct form:

$$V = 0.02 \, \text{m}^3$$
$$\theta = 20°\text{C}, \quad \text{hence} \quad T = 273 + 20 = 293 \, \text{K}$$
$$P = 15 \, \text{MN m}^{-2} = 15 \times 10^6 \, \text{N m}^{-2}$$
$$R = 8.3 \, \text{J mol}^{-1} \text{K}^{-1}$$

From equation [6.10]

$$PV = nRT$$

or

$$15 \times 10^6 \times 0.02 = n \times 8.3 \times 293$$

Rearranging gives

$$n = \frac{15 \times 0.02}{8.3 \times 293} \times 10^6 = 1.23 \times 10^2$$

There are 123 moles of gas.

## GAS TEMPERATURE AND MOLECULAR KINETIC ENERGY

It was stated above that increasing the temperature of a gas, at constant volume, increases the pressure of the gas. This is because the pressure of a gas arises as a result of gas molecules bombarding the walls of the container (see Fig. 6.9). As shown in Fig. 6.9, the collisions of the molecules with the walls are elastic. Thus the molecules rebound with the same *speed* as they are approached. However, there is a change in direction, and hence a change in the velocity on collision. The component of velocity parallel to the wall is unchanged, whereas that perpendicular to the wall is reversed. This results in a change in momentum of the molecules in a direction *perpendicular* to the wall. From Chapter 3 this results in a force exerted on the walls by each molecule during the collision.

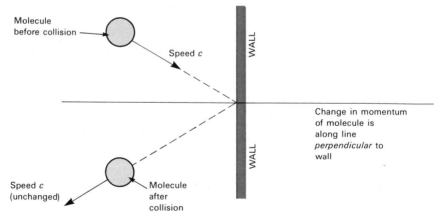

Fig. 6.9   Collision of a gas molecule with the wall of a container

The effect of a single molecule is, of course, very small, and acts for a short time. However, there are very many molecules in a gas, even at very low pressure, and so what is observed is a constant force, at a given temperature, acting on the container walls. This produces the pressure which we observe.

The average translational speed of air molecules at room temperature is around $500 \, \mathrm{m \, s^{-1}}$. When the temperature of the gas is raised the molecules move faster. Hence they exert a larger force during collisions with the walls. Thus the observed pressure rises, as noted above.

The molecules of a gas have a whole range of speeds. Fig. 6.10 shows the number of molecules having speed $c$ at a given temperature. Note also that it shows that the *average* speed of a molecule increases as the temperature of the gas increases.

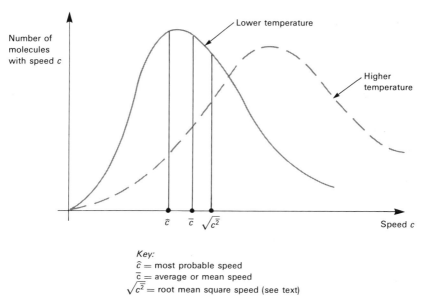

Key:
$\hat{c}$ = most probable speed
$\bar{c}$ = average or mean speed
$\sqrt{\overline{c^2}}$ = root mean square speed (see text)

Fig. 6.10   Distribution of molecular speeds in a gas

It can be shown theoretically that the average kinetic energy of gas molecules is proportional to the absolute temperature of the gas. That is:

$$(\text{Mean KE of gas molecules}) \propto T$$

or

$$\tfrac{1}{2} m \overline{c^2} \propto T$$

where    $m$ = mass of a gas molecule
$\overline{c^2}$ = average or *mean* of the square of the speeds of the molecules
$T$ = absolute temperature

The square root of $\overline{c^2}$, commonly referred to as the root mean square speed (r.m.s. speed, denoted by $\sqrt{\overline{c^2}}$ in Fig. 6.10) is thus of significance. The r.m.s. value is often quoted as a measure of the mean speed of gas molecules.

**EXERCISE 6** _____

*Note.* Standard temperature and pressure (STP) is $0°C$ and a pressure of 76 cm of mercury $(1.013 \times 10^5 \, \text{N m}^{-2})$.

1)  Change the following Celsius temperatures into degrees absolute:

(a)   $0°C$          (b)   $150°C$          (c)   $-30°C$          (d)   $27°C$

2)  Change the following temperatures in degrees absolute into temperatures in degrees Celsius:

(a)   295 K          (b)   113 K          (c)   250 K          (d)   400 K

3)  A gas occupies $200 \, \text{cm}^3$ at $27°C$. What is its volume at $51°C$ if the pressure remains constant?

4)  A gas is held at $47°C$. To what temperature must it be heated to double its volume if the pressure remains constant?

5)  A gas has a pressure of $5 \, \text{N m}^{-2}$ and a temperature of $87°C$. What pressure would the gas exert at $15°C$ if the volume remains constant?

6)  A gas has a volume of $190 \, \text{cm}^3$ at $27°C$ and 80 cm of mercury pressure. What volume would it occupy at STP?

7)  Calculate the mass of air which has a volume of $300 \, \text{cm}^3$ at $15°C$ and 77 cm of mercury pressure if the density of air at STP is $1.29 \, \text{kg m}^{-3}$.

8)  If the density of oxygen is $1.43 \, \text{kg m}^{-3}$ at STP and the mass of 1 mole is 0.032 kg, calculate the molar gas constant $R$.

9)  $0.4 \, \text{m}^3$ of a gas is collected at $37°C$ and at a pressure of 1.2 atmospheres (1 atmosphere $\equiv 1.013 \times 10^5 \, \text{N m}^{-2}$). What volume of gas does this represent at STP? If the molar gas constant $R = 8.3 \, \text{J mol}^{-1}\text{K}^{-1}$, how many moles of gas does this represent?

10)  What mass of helium is present in a volume of $2 \times 10^{-4} \, \text{m}^3$ at $27°C$ and at a pressure of $2 \times 10^5 \, \text{N m}^{-2}$ given that the gas constant $R = 8.3 \, \text{J mol}^{-1}\text{K}^{-1}$ and the mass of 1 mole of helium is 0.004 kg.

# 7

# LONGITUDINAL AND TRANSVERSE WAVES

## WAVE TYPES _____

The most obvious kinds of waves to consider are waves in elastic media since these demonstrate the basis of wave behaviour.

Consider a long spring stretched between two supports. (In a laboratory this might be done on a solid floor.) If the spring is disturbed then the disturbance can be seen to move along the spring — this is a pulse. Two distinct types of pulse are possible (see Fig. 7.1(a) and (b) overleaf).

In (a) the direction of the displacement of the spring is in the direction of pulse movement. The pulse is a compression — rarefaction situation.

Notice the direction of the displacement is completely fixed.

In (b) the displacement is at right angles to the wave direction.

Although the direction of displacement in this case is fixed it could have taken any direction on the plane indicated and the situation would have been effectively the same.

The pulse in (a) is *longitudinal.*

The pulse in (b) is *transverse.*

Let us now look in more detail at the reason for the movement of the pulses in the spring. The spring represents any elastic medium (i.e. any medium where an attempt to reduce its dimensions results in a proportional change in the forces necessary to cause such a reduction). This, therefore, could be an elastic solid or a liquid (not a system one normally tries to compress) or a gas.

The situations (a) and (b) have similarities but let us deal with them separately.

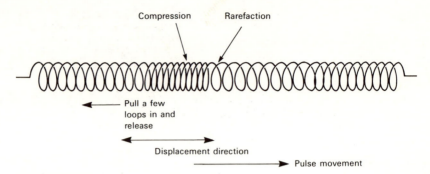

Compression    Rarefaction

Pull a few
loops in and
release

Displacement direction

Pulse movement

(a) Longitudinal pulse

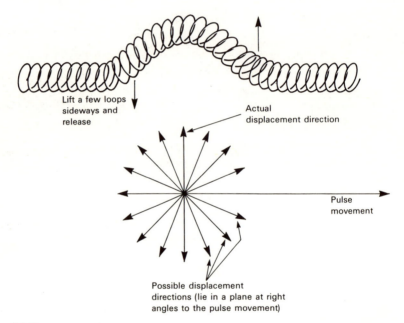

Lift a few loops
sideways and
release

Actual
displacement direction

Pulse
movement

Possible displacement
directions (lie in a plane at right
angles to the pulse movement)

(b) Transverse pulse

Fig. 7.1    Waves in a spring

## LONGITUDINAL PULSE

This situation is best described with reference to Fig. 7.2(a). The disturbed portion of the spring is only a small section of the total length. At the front of this pulse the loops crowd together in compression forcing sideways motion ahead. This crowding is compensated by a region, behind the most crowded region, of slightly greater separation than normal. Here the loops settle back to an undisturbed situation causing a rarefaction.

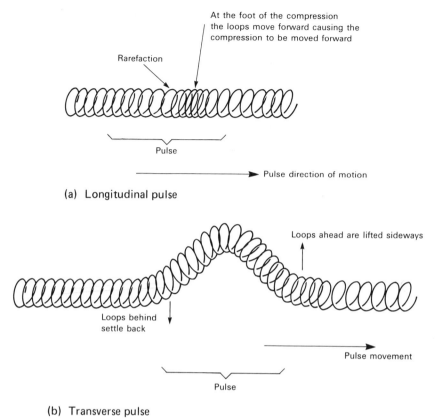

Fig. 7.2   Pulse motion in a spring

## TRANSVERSE PULSE

In Fig. 7.2(b) the spring above the main body pulls it sideways propagating the motion forward. Meanwhile the loops settle back to normality at the rear of the pulse.

These situations are pulses. If however, the disturbance is repetitive then the result is a wave where the repetition length, $\lambda$, is called the *wavelength* (see Fig. 7.3 overleaf).

In parts (a) and (b) of Fig. 7.3 the situations on the wave are shown at some instant in time. The figures are effectively photographs of what is happening along the spring. There is another way of displaying the situation and that is to show what is happening to one loop as time goes on (Fig. 7.3(c)). With reference to Fig. 7.3(c), then

$$T = \frac{1}{f} \qquad\qquad [7.1]$$

where    $T$ = the periodic time (time for one complete oscillation) in seconds (s)

$f$ = the frequency (number of oscillations in one second) in hertz (Hz)

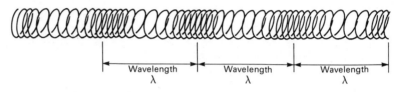

(a)  Longitudinal wave

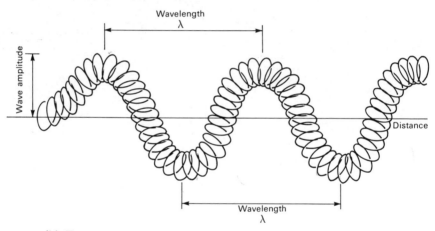

(b)  Transverse wave

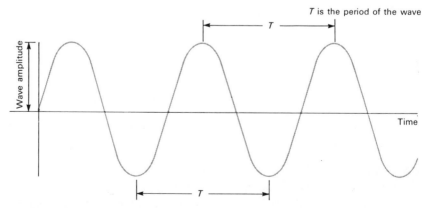

(c)  Motion of a given loop

Fig. 7.3    Wave motion

Now in $T$ seconds the wave moves forward a distance $\lambda$. Hence the velocity $c$ of the wave is given by

$$c = \frac{\lambda}{T}$$

From equation [7.1], since $f = \frac{1}{T}$, then

$$c = f\lambda \qquad \qquad [7.2]$$

This relationship holds whatever kind of wave we are considering.

Examples of different kinds of waves are:

*Longitudinal waves* — sound waves in solids, liquids or gases.
*Transverse waves* — (a) light, X-rays, $\gamma$-rays, infra-red, etc.
(b) water waves on the surface of a liquid.

## WORKED EXAMPLE 1

The period of a wave is $2 \times 10^{-6}$s. What is the frequency?

*Solution*

From equation [7.1] we have

$$T = \frac{1}{f}$$

$$2 \times 10^{-6} = \frac{1}{f}$$

Rearranging gives

$$f = \frac{1}{2 \times 10^{-6}} = 5 \times 10^5 = 500 \times 10^3$$

or $\qquad f = 500\,kHz$

## WORKED EXAMPLE 2

Light travels at a speed of $3.00 \times 10^8\,m\,s^{-1}$. For a wave of frequency $500\,kHz$, what would be the wavelength?

*Solution*

From equation [7.2]

$$c = f\lambda$$

$$3.00 \times 10^8 = 500 \times 10^3 \times \lambda$$

or
$$\lambda = \frac{3 \times 10^8}{5 \times 10^5} = 600$$

Hence wavelength = 600 m.

Notice the unit of wavelength is the metre (m). It is useful at this point to remember the subdivisions and multiples of basis units:

m before a symbol $= 10^{-3}$
$\mu$ before a symbol $= 10^{-6}$
n before a symbol $= 10^{-9}$
k before a symbol $= 10^{+3}$
M before a symbol $= 10^{+6}$

# POLARISATION

Previously it was mentioned that the direction of displacement is completely specified for a longitudinal wave once the direction of wave motion is known. In the case of a transverse wave this was not so; there was still any direction possible, within 360°. Where a transverse wave *is* limited to one direction only, i.e. where the displacement direction and the direction of motion define a plane, that particular transverse wave is said to be *plane polarised*. This can be obtained from a range of disturbances by a polariser.

## POLARISATION OF LIGHT

Ordinary light is unpolarised. It can be plane-polarised by passage through a sheet of Polaroid® — see Fig. 7.4. Only vibrations parallel to the transmission axis of this Polaroid (referred to as the *polariser*) are transmitted. This arises because of the structure of Polaroid. Light incident on Polaroid is split into components with vibrations parallel and perpendicular to the transmission axis. The vibration component perpendicular to the transmission axis is absorbed whilst that parallel to the axis is transmitted. This effect is called *dichroism*.

We can show that the light emerging from the polariser is plane-polarised by using a second sheet of Polaroid (referred to as the analyser) as shown in Fig. 7.5. If the analyser is rotated there will be some position where no light gets through it. This occurs when the transmission axes of polariser and analyser are at 90° and are referred to as 'crossed' (see Fig. 7.5(a)).

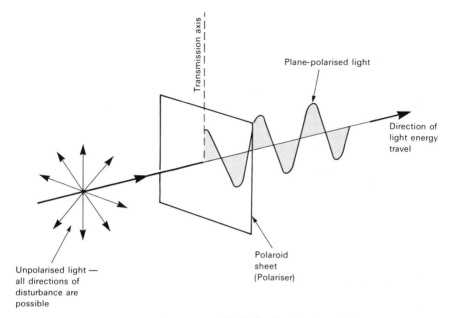

Transmission axis

Plane-polarised light

Direction of
light energy
travel

Unpolarised light —
all directions of
disturbance are
possible

Polaroid
sheet
(Polariser)

Fig. 7.4  Production of plane-polarised light

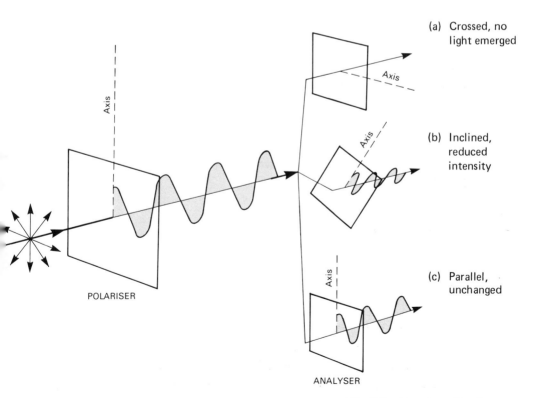

(a)  Crossed, no
light emerged

Axis

(b)  Inclined,
reduced
intensity

Axis

Axis

Axis

POLARISER

(c)  Parallel,
unchanged

Axis

ANALYSER

Fig. 7.5  Rotation of analyser

When the polariser and analyser axes are parallel, as in Fig. 7.5(c), the intensity transmitted is a maximum and equals the intensity transmitted by the polariser.

At intermediate positions between the crossed and parallel states (see Fig. 7.5(b)) the transmitted intensity is intermediate between zero and that transmitted by the polariser. This is because the analyser transmits only the component of the incident disburbances which are parallel to its transmission axis. The variation of transmitted intensity with angle $\theta$ between transmission axes of polariser and analyser is shown in Fig. 7.6.

Only a transverse wave could be polarised and thus these experiments demonstrate that light is a transverse wave.

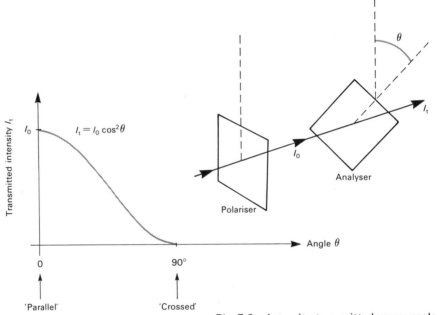

Fig. 7.6    Intensity transmitted versus angle $\theta$

## POLARIMETRY AND SACCHARIMETRY

When plane-polarised light is passed through solutions of some substances the plane of polarisation is rotated — see Fig. 7.7(a). We say these solutions are *optically active*. Sugars are in this category since they consist of corkscrew like molecules. Substances which rotate the plane of polarisation to the 'right' (see Fig. 7.7(b)) are called *dextro*rotary. Those which rotate it in the opposite direction are called *laevo*rotary — they are rarely found in nature.

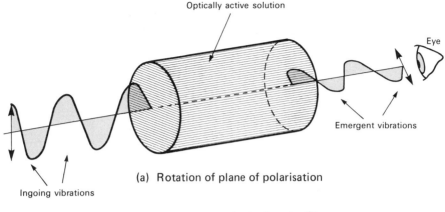

(a) Rotation of plane of polarisation

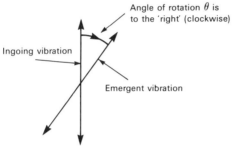

(b) Rotation as seen by the eye in (a)

Fig. 7.7   Optical activity

The angle of rotation depends upon the type of substance (e.g. sugar type), its concentration $C$ and the path length $l$ transversed by the light beam. It is found that

$$\theta = \alpha l C \qquad [7.3]$$

where    $\theta$ = angle of rotation
         $\alpha$ = specific optical rotation
         $l$ = length of solution transversed by light beam
         $C$ = concentration of substance in solution

Note that $\alpha$ is a constant for a given substance in solution at a given temperature and for a particular wavelength of incident light.

The effect provides a rapid and accurate method for determining the concentration of a sugar in solution. A device known as a *polarimeter* is used and is shown in schematic form in Fig. 7.8. It consists of a polariser and analyser between which is placed a tube of fixed length $l$. The polariser and analyser are crossed when the tube is full of distilled water and the position of the analyser noted. A solution of known concentration $C$ is now placed

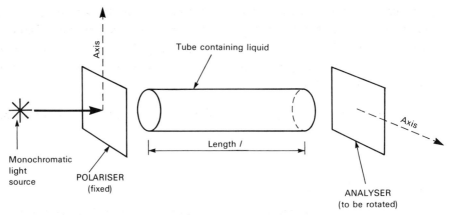

Fig. 7.8    A polarimeter — schematic form

in the tube and the analyser rotated until the transmitted light intensity is again zero. The angle of rotation $\theta$ is noted. The experiment is repeated for a range of known concentrations and $\theta$ noted in each case. A graph of $\theta$ versus $C$ is drawn. According to equation [7.3] this should be linear with gradient $\alpha l$.

A solution of unknown concentration $C'$ is now placed in the tube and the angle of rotation $\theta'$ noted. Hence $C'$ is found from the graph, or by use of equation [7.3] if gradient $\alpha l$ of graph is known.

A diagram of a simple polarimeter is shown in Fig. 7.9. The polariser and analyser may be polaroid or, more likely, specially cut calcite crystals called *Nicol prisms* which are more effective than Polaroid. The analyser can be

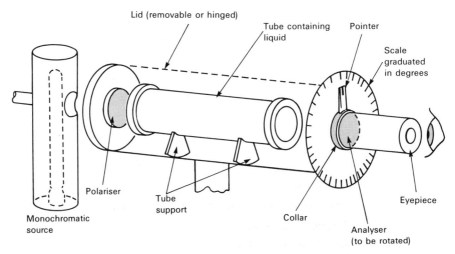

Fig. 7.9    A simple polarimeter

rotated and is mounted in a collar carrying a pointer which moves over a circular scale graduated in degrees. (In some instruments the polariser may also be rotated to enable the initial reading of the analyser to be set at zero.) Lenses may be present in the instrument so that light passes through the solution as a parallel beam.

Some instruments are specially designed to measure sugar concentrations directly and are termed *saccharimeters.*

### WORKED EXAMPLE 3

In an experiment using a polarimeter illuminated with sodium light, known concentrations of cane sugar in distilled water were used. The tube length was 0.20 m. The following results were observed:

| Concentration $C$ of sugar/g cm$^{-3}$ | 0.20 | 0.10 | 0.05 |
|---|---|---|---|
| Angle of rotation $\theta$/degrees | 27 | 13 | 7 |

(a) Plot a graph of $\theta$ versus $C$ and estimate the specific optical rotation of this sugar in water.

(b) Find the concentration of sugar which rotates the plane of polarisation through 20°.

### Solution

(a) The graph is shown in Fig. 7.10. From equation [7.3] it is a straight line with gradient $\alpha l$. Note it passes through the point (0.20, 27)

  (i) *SI units*:

   We note that $1.0 \text{ g cm}^{-3} \equiv 10^3 \text{ kg m}^{-3}$

$$\text{Gradient of graph} = \frac{27}{0.20 \times 10^3}$$

$$= 0.135 \text{ deg kg}^{-1}\text{m}^3$$

   Since $l = 0.20 \text{ m}$ and gradient is $\alpha l$, then

$$\alpha = \frac{\text{Gradient}}{l} = \frac{0.135}{0.2}$$

$$= 0.675 \text{ deg kg}^{-1}\text{m}^2$$

  (ii) *Polarimetry units*:

   The units traditionally used in polarimetry are as follows:

   Rotation $\theta$ in degrees
   Length $l$ in decimetres
   Concentration $C$ in g cm$^{-3}$

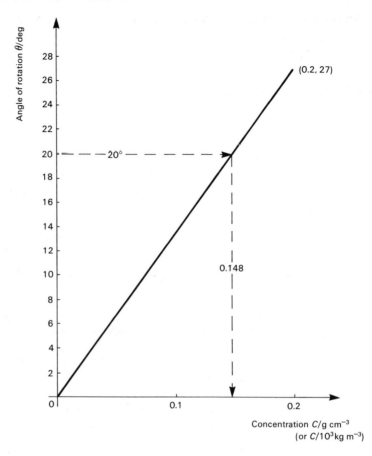

Fig. 7.10    Solution for Worked Example 3

Working in these units, we have

$$\text{Gradient of graph} = \frac{27}{0.20} = 135$$

Since $l = 2.0$ and gradient is $\alpha l$ then

$$\alpha = \frac{\text{Gradient}}{l} = \frac{135}{2}$$

$$= 67.5$$

We say the specific optical rotation of cane sugar is 67.5 degrees. This is the figure which is often quoted by suppliers.

(b)    From the graph of Fig. 7.10, $\theta = 20°$ corresponds to a concentration $C$ of $0.148\,\text{g cm}^{-3}$ ($0.148 \times 10^3\,\text{kg m}^{-3}$).

Alternatively, using polarimetry units, we have

$$\theta' = 20°$$
$$\alpha = 67.5$$
$$I = 2.0$$

From equation [7.3]

$$C' = \frac{\theta'}{\alpha I} = \frac{20}{(67.5 \times 2.0)} = 0.148$$

The unknown concentration $C' = 0.148 \, \mathrm{g \, cm^{-3}}$.

In fact, due to the difficulties involved in measuring $\theta$ accurately, this experiment is unlikely to yield results more precise than two significant figures. So we write $C' = 0.15 \, \mathrm{g \, cm^{-3}}$.

**EXERCISE 7** _____

1)   A tuning fork vibrates with  a period of 2.32 ms. To what frequency does this correspond?

2)   Radio 1 is at 1053 kHz. What wavelength does this correspond to? (Radio waves, like all electromagnetic waves, travel at $3 \times 10^8 \, \mathrm{m \, s^{-1}}$.)

3)   At a frequency of 2 kHz, the wavelength of sound in air is 15 cm. What is the velocity of sound in air?

4)   The wavelength of mercury green light is given as 555 nm. What frequency does this correspond to if the velocity of light is $3 \times 10^8 \, \mathrm{m \, s^{-1}}$.

5)   Find the concentration of cane sugar in a liquid of which a column, 20 cm long rotates the plane of polarisation of sodium light through 30° (specific rotation of cane sugar is 67.5°).

6)   The rotation observed when using a polarimeter of tube length 20 cm was 8.5° for a concentration of a sugar of 0.05 g cm⁻³. Calculate a value for the specific optical rotation of the sugar.

# INTERFERENCE, DIFFRACTION AND SPECTRA

## INTRODUCTION

Interference and diffraction of waves are phenomena that happen under quite specific situations. Interference is the interaction of waves crossing the same region of space. Diffraction is the effect on a single wave of restricting the wavefront. Interference and diffraction are of interest directly to a physicist. For a chemist the interest is to understand the workings of spectrometers because these enable the chemist to analyse samples of material.

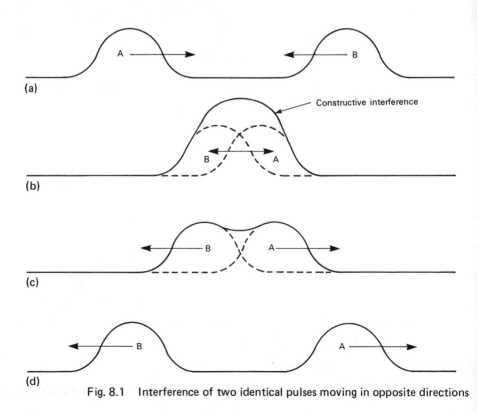

Fig. 8.1   Interference of two identical pulses moving in opposite directions

# INTERFERENCE

## PULSES

To explain what happens when waves of the same type meet, it is best first to consider similar pulses. In Fig. 8.1, a description is given of what can be demonstrated with a stretched spring. For ease of understanding the spring will be shown as a single line. Initially, in Fig. 8.1(a), the pulses are approaching one another; they meet in Fig. 8.1(b); start to be seen as separate entities in Fig. 8.1(c), and finally are seen separating in Fig. 8.1(d). When they were in the same region of the spring what was seen was the *algebraic sum* of the two displacements. 'Algebraic' means taking account of whether the displacement was up (positive) or down (negative). This becomes clearer in Fig. 8.2 where the pulses have displacements in opposite directions. The effect is described as follows. Pulses A and B interfere with one another.

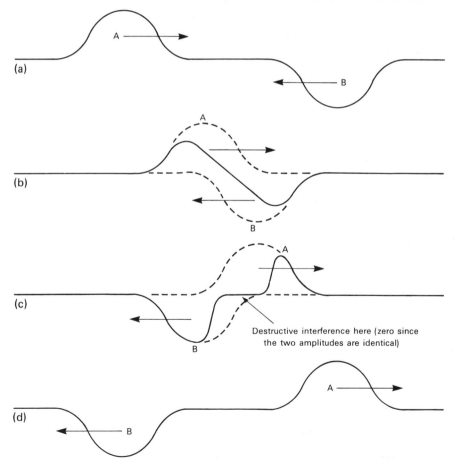

Fig. 8.2    Interference of positive and negative pulses moving in opposite directions

When they interfere the total situation seen is calculated with the aid of the principle of superposition, i.e. that the result is the algebraic sum of the two.

## WAVES

The situation becomes more difficult to describe with waves because of their repetitive nature, and it is this repetition which imposes certain restrictions:

(a)    The waves must have the same wavelength for interference to take place.

(b)    For a good interference situation the amplitude of the waves must be roughly the same. (In the case of Figs. 8.1 and 8.2 we made their amplitudes equal which is the easiest case to consider.)

Consider the simplest case of two sources of water waves on a pond producing waves of the same amplitude and frequency. In order to display what is effectively a three-dimensional situation in only two dimensions (i.e. on paper), we have to introduce some terminology. Fig. 8.3 shows how wave motion is represented for one source.

Now it is possible to represent three-dimensional movement more easily, let us see what happens for the two sources of water waves mentioned, Fig. 8.4.

Along line XX, bisecting the line between the sources, wave crests from source P arrive with crests from source Q. The result is shown in Fig. 8.5(a), and corresponds to a wave of twice the amplitude. We say *constructive interference* has taken place. This is also the case along the other lines of constructive interference (anti-nodal lines) in Fig. 8.4. Along the nodal lines, however, the situation is entirely different. Fig. 8.5(b) shows that here a crest from P arrives with a trough from Q, and vice versa. The result is zero displacement with time — that is, no disturbance at all. This is termed *destructive interference.*

To summarise:

(a)    Where the path difference between the wave from P and the wave from Q is a whole number of wavelengths constructive interference occurs to give a double amplitude disturbance.

(b)    Where the path difference is an odd number of half wavelengths the result is no disturbance with time at all (destructive interference).

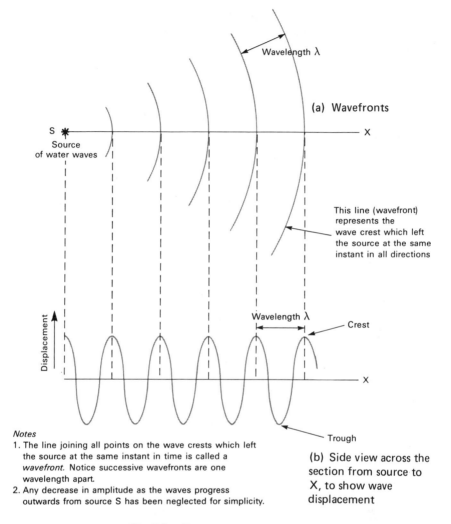

(a) Wavefronts

This line (wavefront) represents the wave crest which left the source at the same instant in all directions

Wavelength λ

Crest

Trough

**Notes**
1. The line joining all points on the wave crests which left the source at the same instant in time is called a *wavefront*. Notice successive wavefronts are one wavelength apart.
2. Any decrease in amplitude as the waves progress outwards from source S has been neglected for simplicity.

(b) Side view across the section from source to X, to show wave displacement

Fig. 8.3    Representation of wave motion from a single source

The actual appearance of the pattern depends on the ratio of the source separation to the wavelengths of the wave (Fig. 8.6).

With reference to Fig. 8.6:

(a)    As the wavelength shortens (part (b)) we must go a shorter distance from the centre line before a path difference of one wavelength is reached.

(b)    As the source separation decreases (part (c)) we must go a larger distance from the centre line before a position of one wavelength path difference is reached.

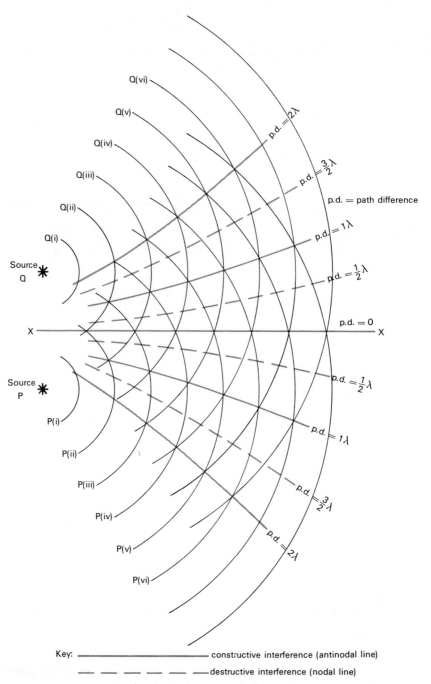

Key: ————————————————— constructive interference (antinodal line)

          — — — — — — — destructive interference (nodal line)

Fig. 8.4    Two-source interference pattern

(a)

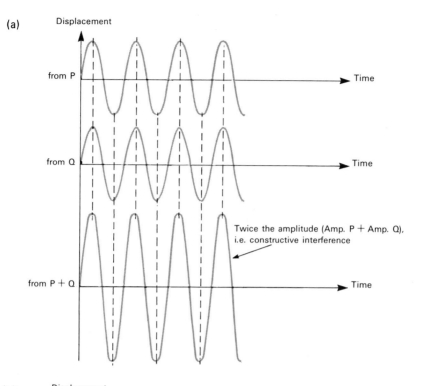

(b)

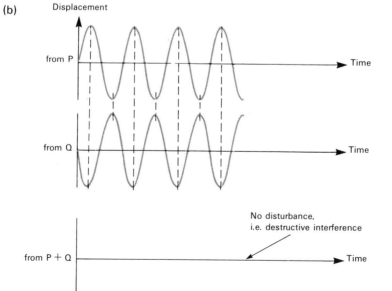

Fig. 8.5  Wave displacements:  (a) along a line of *constructive* interference,
(b) along a line of *destructive* interference

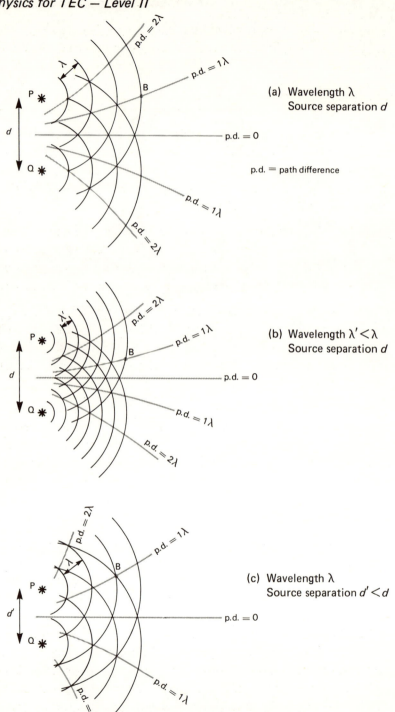

(a) Wavelength λ
    Source separation d

p.d. = path difference

(b) Wavelength λ' < λ
    Source separation d

(c) Wavelength λ
    Source separation d' < d

Fig. 8.6    Appearance of pattern as wavelength and source separation changes
(QB − PB = one wavelength)

Thus the pattern contracts as the wavelength shortens (for a given source separation), and expands as the source separation decreases (for a given wavelength).

# DIFFRACTION

For the simple case of waves spreading out from a point source all that is needed to show the positions of the wavefronts at any instant is a set of compasses. These are set to 1, 2, etc. times the wavelength. The result is a set of concentric circles. Parts of these are shown to the left of the aperture in Fig. 8.7(a).

(An alternative method of finding the position of the next wavefront ahead is to set the compasses to *one* wavelength, and use as centre of this circle all the points of the present wavefront. Only the forward arc of a circle is drawn, and the wavefront lies along the line which ends up darkest.)

Where the wavefront is restricted in any way the result is the appearance of wave energy in a region of geometrical shadow (Fig. 8.7(a)). The resulting phenomenon is called diffraction of waves. It occurs for all types of wave, whether longitudinal or transverse. It also occurs for all types of transverse waves from water waves to radio waves.

The diffracted waves are simple if the restriction is small, because the restriction acts as a secondary source, and can for all intents and purposes be treated as a primary source of waves. Where the restriction is not small but wide (here 'wide' means greater than one wavelength in width), then the wave pattern becomes more complex as waves originating from different parts of the aperture interfere with one another (Fig. 8.7(b) and (c)). The result is alternate directions of high wave energy and zero wave energy.

Diffraction is especially important, because although for water waves and microwaves it is easy to get sources that are in phase, or have some small phase difference, for light this is not so. This is because ordinary light is emitted by excited atoms, and the emission occurs for very small intervals of time. There is a random time interval between emissions. Thus two different optical sources would not have a constant phase difference between them. Hence an interference pattern would not be visible to the eye. We say that two different optical sources are *incoherent*. In practice, in order to get a visible interference pattern using light the two sources must originate, *via diffraction*, from the *same original source*. One method of doing this is by diffraction. This is described in Young's double-slit experiment (page 109).

In a laser *stimulated emission* takes place and all atoms emit in phase.

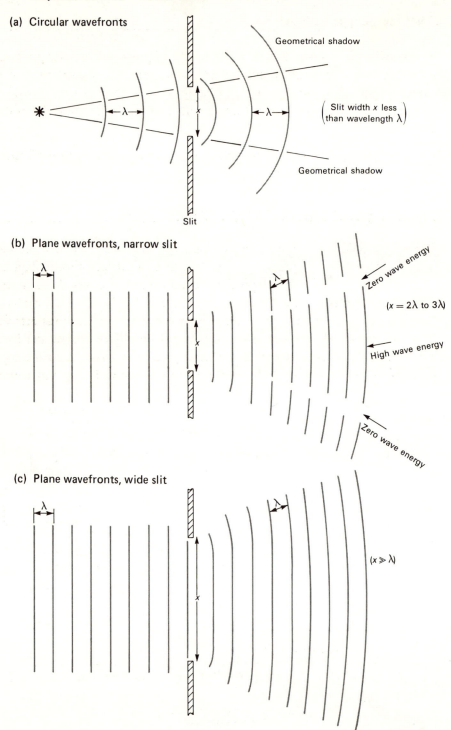

(a) Circular wavefronts

Geometrical shadow

Geometrical shadow

$\left(\begin{array}{c}\text{Slit width } x \text{ less}\\ \text{than wavelength } \lambda\end{array}\right)$

Slit

(b) Plane wavefronts, narrow slit

Zero wave energy

$(x = 2\lambda \text{ to } 3\lambda)$

High wave energy

Zero wave energy

(c) Plane wavefronts, wide slit

$(x \gg \lambda)$

Fig. 8.7   Diffraction causes waves to spread into a region of geometrical shadow

# THE NATURE OF LIGHT

Historically the nature of light was a source of great debate: was it wavelike or corpuscular? The answer changed at various periods in history until at this present time a physicist would probably answer that if you looked for particle behaviour you would find it, and that the same could be said for wave behaviour.

One experiment that shows beyond doubt the *wavelike* nature of light is Young's double-slit experiment.

# YOUNG'S DOUBLE-SLIT EXPERIMENT

Experiments to show interference of light are more difficult than for water waves because, as described above, each atom of a source emitting light does so in random bursts. Thus light from a source is not as regular as a water wave source. So, in order to show interference for light waves, therefore, light from the same source travelling by two different paths must interfere. This is achieved by allowing the light to fall on adjacent slits, and letting the resulting *diffraction* that occurs throw the light across the same region of space. The result is an *interference pattern* as shown in Fig. 8.8, which is

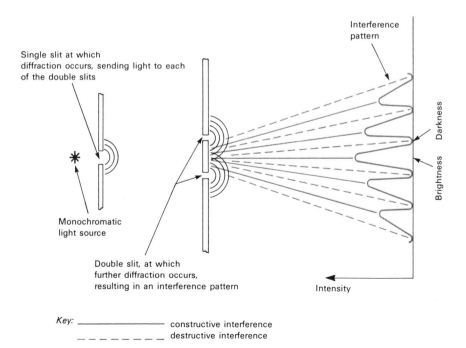

Fig. 8.8 Interference pattern using Young's double-slit arrangement

similar to that of Fig. 8.4 for water waves. This pattern of light can only be due to the *wave nature* of light.

The diffraction of the light at each slit can be observed if the other is covered. The result can show all the effects of Fig. 8.7 as the slit width is altered.

Instead of light of a single wavelength (colour) a white light source may be used. Let us consider just the three primary colours (red, green and blue) for the Young's double-slit experiment. The result is Fig. 8.9. This occurs because, as indicated in Fig. 8.6, the different wavelengths have interference patterns which are spaced differently. The larger the wavelength, then the more widely spaced is the interference pattern. However, all wavelengths produce a maximum along the centre line (XX in Fig. 8.4). This means that, as shown in Fig. 8.9, the centre would be white, and a mixed-up rainbow effect would result on either side. *Dispersion* of the light has occurred. The light has been split into its component colours or wavelengths.

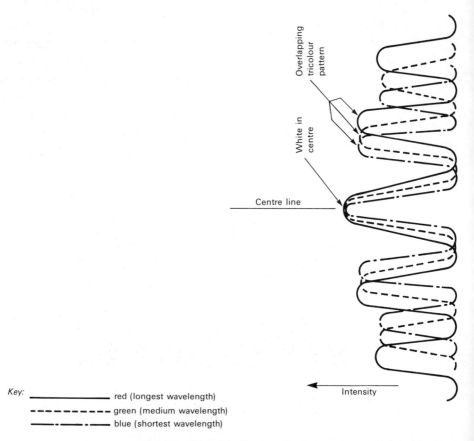

Fig. 8.9   Young's slits interference pattern using three colours

This dispersion of light has practical uses. Light from different elements contains different wavelengths, so once these wavelengths are identified then the chemical composition of the light source can be identified. Alternatively, it has been found that an element that emits a wavelength also absorbs it. Hence it is possible to shine white light through a material and identify its constituents by the missing (absorbed) wavelengths. These two ideas are the basis of *emission spectrometry* and *absorption spectrometry*. For practical purposes the dispersion of light in Young's double-slit experiment is not efficient enough. The situation is better if more slits are employed. A device with thousands of equal slits regularly spaced is called a *diffraction grating*; 5000 lines per cm is commonplace for such a grating.

## THE DIFFRACTION GRATING

A diffraction grating consists of a large number of extremely narrow slits. It can be made by ruling many parallel equidistant lines on a glass plate (see Fig. 8.10).

The effect of many fine slits very close together changes the pattern in three ways from the Young's double-slit situation. First, since the slits are very much narrower, the diffraction is much greater, i.e. more light is given off at greater angles. Secondly, the total area through which light passes is increased (because of the large number of slits), and hence the interference beams are more intense. Finally, because interference occurs between all these slits, the line width in the interference pattern is narrower. This final point means

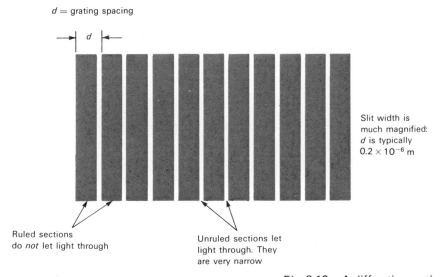

$d$ = grating spacing

Slit width is much magnified: $d$ is typically $0.2 \times 10^{-6}$ m

Ruled sections do *not* let light through

Unruled sections let light through. They are very narrow

Fig. 8.10   A diffraction grating

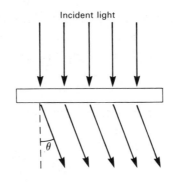

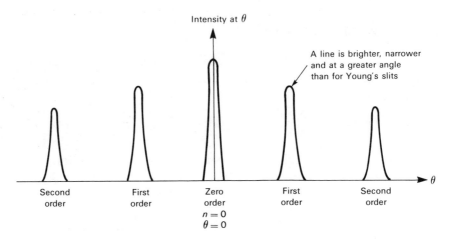

Fig. 8.11    Appearance of maxima produced by a diffraction grating

positions of maxima are more precisely located. Fig. 8.11 shows the pattern produced by a diffraction grating (when parallel incident light is used). The order number refers to the number of wavelengths path difference between light from adjacent slits.

In Fig. 8.12(a) just four adjacent slits of a grating are shown. (This is for the sake of clarity — in a real grating there are thousands of slits.) The distance between the centres of successive slits is called the *grating spacing d*. A parallel incident beam of wavelength λ is incident normally on the grating. A parallel diffracted beam, at some angle θ with the normal to the grating is brought to a focus by the telescope lens. This produces an image of the source slit.

The path difference between adjacent parallel rays is, from Fig. 8.12(b), equal to $d \sin \theta$. For constructive interference between light waves from adjacent slits the path difference must be equal to a whole number of wavelengths. That is,

$$d \sin \theta = n\lambda \qquad [8.1]$$

where $n = 0, 1, 2, 3 \ldots$

Thus a diffracted beam is seen only in those directions for which angle $\theta$ is given, from equation [8.1] by

$$\theta = \sin^{-1}\left(\frac{n\lambda}{d}\right) \qquad [8.2]$$

This means that intense light beams are given off at various angles according to their wavelength. Note that as wavelength increases, so too does angle $\theta$ (for a given value of $n$).

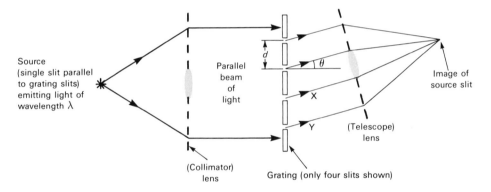

(a) Parallel beams passing through (part of) a diffraction grating

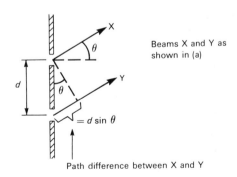

(b) Path difference between adjacent rays $= d\sin\theta$

Fig. 8.12 Formation of an image using a diffraction grating set-up

## WORKED EXAMPLE 1

Light of wavelength 550 nm (green) is incident normally on a diffraction grating which has 5000 lines per cm. At what angles do interference maxima emerge from the grating?

*Solution*

5000 lines per cm means

$$\text{Grating spacing } d = \frac{1}{5000}\text{ cm}$$

Rearranging equation [8.1] gives

$$\sin\theta = \frac{n\lambda}{d}$$

where

$$\lambda = 550 \times 10^{-9}\text{ m}$$

$$d = \frac{1}{5000}\text{ cm} = 2.00 \times 10^{-6}\text{ m}$$

and $n = 0, 1, 2, 3, 4 \ldots$

Thus

$$\sin\theta = \frac{n \times 550 \times 10^{-9}}{2.00 \times 10^{-6}}$$

or

$$\sin\theta = n \times 0.275$$

For $n = 1$, the first-order maximum,

$$\sin\theta_1 = 0.275$$

or

$$\theta_1 = 15°58'$$

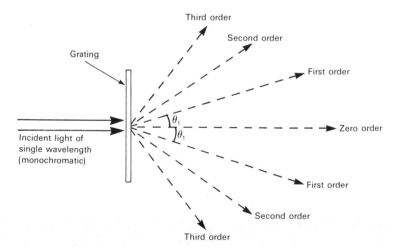

Fig. 8.13    Angular distribution of various orders for a single wavelength incident on a diffraction grating

For $n = 2$, the second-order maximum,

$$\sin \theta_2 = 0.550$$

or
$$\theta_2 = 33°22'$$

For $n = 3$, the third-order maximum

$$\sin \theta_3 = 0.825$$

or
$$\theta_3 = 55°35'$$

In this case $n = 4$, the fourth order, does not occur since $4 \times 0.275 = 1.10$ and sine values only go up to 1.00. Note that for $n = 0$ then $\theta = 0$. So in this case there would be the straight-through zero-order line, and three lines on each side. This is shown in Fig. 8.13 in schematic form. Note also that the high-angle orders are faint, and so are not of much practical use.

## WORKED EXAMPLE 2

A grating has 6000 lines per cm. Calculate the angular separation of wavelengths 650 nm, (red) and 500 nm (blue) in the first order.

*Solution*

Grating spacing $d = \dfrac{1}{6000} \text{cm} = \dfrac{1}{6} \times 10^{-5} \text{m}$

Rearranging equation [8.1] gives

$$\sin \theta = \frac{n\lambda}{d}$$

where
$$n = 1$$

$$d = \frac{1}{6} \times 10^{-5} \text{m}$$

For
$$\lambda = 650 \times 10^{-9} \text{m}, \quad \text{let} \quad \theta = \theta_1$$

So
$$\sin \theta_1 = \frac{1 \times 650 \times 10^{-9}}{1/6 \times 10} = 0.390$$

or
$$\theta_1 = 22°57'$$

For
$$\lambda' = 500 \times 10^{-9} \text{m}, \quad \text{let} \quad \theta = \theta_1'$$

so
$$\sin \theta_1' = \frac{1 \times 500 \times 10^{-9}}{1/6 \times 10^{-5}} = 0.300$$

or
$$\theta_1' = 17°28'$$

$$\theta_1 - \theta_1' = 5°29'$$

The angular separation in the first order is $5°29'$. Note that *larger wavelengths emerge at larger angles*. By repeating the calculation for higher orders we can also show that the *angular separation of two wavelengths increases as the order increases*. These features are shown in Fig. 8.14.

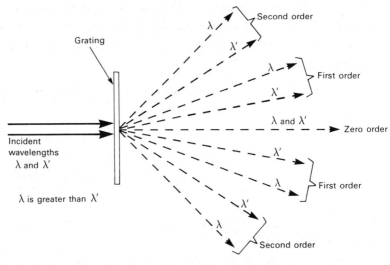

Fig. 8.14    Angular distribution of various orders for two wavelengths incident on a diffraction grating

## USE OF DIFFRACTION GRATING TO MEASURE WAVELENGTH

Equation [8.1] shows that once $\theta$, $d$ and $n$ are known, the wavelength $\lambda$ may be measured. As shown in Fig. 8.11 the angular width of the maxima are small, and this enables $\theta$ to be measured accurately. Since $d$ and $n$ are also accurately known the wavelengths may be measured accurately.

A diagram of a diffraction grating on a spectrometer is shown in Fig. 8.15. The light source, with unknown wavelength(s), illuminates the slit of the collimator. When the slit is at the focus of the collimator lens a parallel

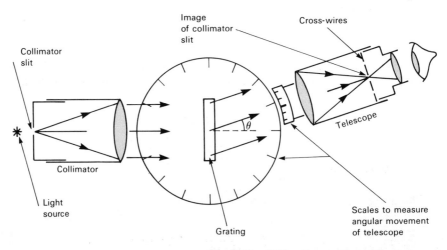

Fig. 8.15    Diffraction grating and spectrometer

beam of light is incident on the grating. The telescope receives the diffracted light as parallel beams. These are focused at the crosswires of the telescope, and form images of the collimator slit. The angle $\theta$ at which a beam emerges is measured by a circular scale.

The following adjustments should be made before using the instrument:

(a)  The collimator to emit parallel light.

(b)  The telescope to receive and focus parallel light.

(c)  The lines on the grating to be vertical.

(d)  The plane of the grating to be perpendicular to the axis of the collimator (i.e. incident light to be normal to grating).

The procedures for ensuring correct adjustment are lengthy, and are described in laboratory textbooks.

The telescope is now rotated to view the diffracted images at the centre of the cross wires. The order $n$ of the image, and its angle $\theta$ from the undeviated position is noted in each case. Hence the wavelength(s) emitted by the source are found by rearranging equation [8.1] to give

$$\lambda = \frac{d \sin \theta}{n} \qquad [8.3]$$

## WORKED EXAMPLE 3

Monochromatic light is incident normally on a diffraction grating with $5.000 \times 10^5$ lines per metre. The two first-order diffraction maxima are separated by an angle of 37°56′ (see Fig. 8.16), and the two second-order maxima by 81°4′. Calculate the wavelength of the incident light.

## Solution

The angles given in Fig. 8.16 are twice the angles from the straight through position. (In practice this is done to offset any error in setting the grating normal to the incident beam.)

Hence $\qquad\qquad \theta_1 = \frac{1}{2}$ of 37°56′ = 18°58′

and $\qquad\qquad \theta_2 = \frac{1}{2}$ of 81°4′ = 40°32′

From equation [8.3]

$$\lambda = \frac{d \sin \theta}{n}$$

where $\qquad\qquad d = \frac{1}{5 \times 10^5} = 0.2 \times 10^{-5}\,\text{m}$

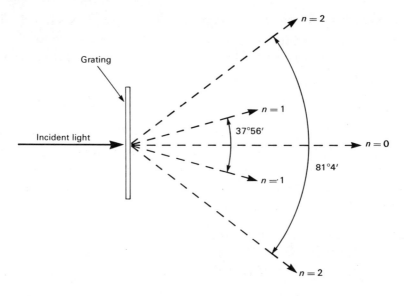

Fig. 8.16    Information for Worked Example 3

For                    $n = 1, \quad \theta = 18°58'$

Hence                  $\lambda = \dfrac{d \sin \theta_1}{n_1} = \dfrac{0.2 \times 10^{-5} \times 0.3250}{1}$

$= 0.6500 \times 10^{-6} \, m$

For                    $n = 2, \theta_2 = 40°32'$

Hence                  $\lambda = \dfrac{d \sin \theta_2}{n_2} = \dfrac{0.2 \times 10^{-5} \times 0.6498}{2}$

$= 0.6498 \times 10^{-6} \, m.$

The average wavelength is $0.6499 \times 10^{-6} \, m.$

## EMISSION SPECTROSCOPY

In the industrial laboratory situation the diffraction grating can be used in conjunction with a spectrometer to identify the elements present in a light source by finding out the emitted wavelengths. Once the wavelengths are known these can be compared with data which give the intense lines for each element of the periodic table. This type of spectroscopy is called *emission spectroscopy* because the light emitted by a source is being studied.

The source is one of the following:

(a)    flame into which the material is introduced;

(b)    discharge tube;

(c)    electric arc;

(d)    electric spark.

The difference between the last two is only in the size of the currents and voltages. In an arc the voltage is low, and the current is high, whereas in a spark the current is low and the voltage is high. All of these sources completely atomise the material giving line spectra. Such techniques are further discussed in the next section.

## DISPERSION AND SPECTROMETRY

*Dispersion* means 'splitting up'. In this context light of many colours (e.g. white·light), is split up so that different wavelengths emerge in different directions. A diffraction grating disperses white light. Another method of effecting a separation of wavelengths is through *refraction* (Fig. 8.17). Notice that at the interface there is dispersion. This would be cancelled out if the second boundary were parallel to the first, but if a prism of glass is used then the separation is increased (Fig. 8.18). The amount of dispersion depends on the type of glass from which the prism is made, and also on the angle $A$ of the prism.

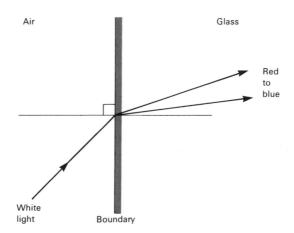

Fig. 8.17    Dispersion at an air–glass interface (the angle between red and blue is exaggerated)

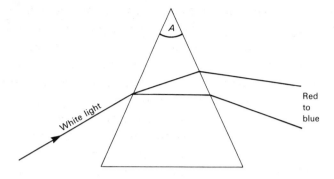

Fig. 8.18    Dispersion by a prism

White light is light composed of many colours or wavelengths from about 400 nm (blue) to 700 nm (red). This is the visible range of the electro-magnetic spectrum (see Chapter 9). Light covering the whole of this range is emitted by very hot solid objects (e.g. a tungsten lamp). The amount of light emitted from such a source at any particular wavelength is mainly a function of the temperature of the source (i.e. the hotter the source the more light is emitted at every wavelength). Notice it does not depend on the material of the source at all.

If the atoms emitting the light are widely separated, as in a gas, then the emitted light is of a totally different composition, dependent on the atoms

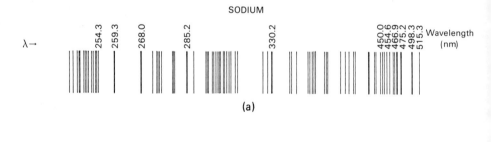

Fig. 8.19    Spectrogram of (a) sodium and (b) lithium

involved. Now the light given out depends on the element to which the atoms belong — the light is characteristic of the element. The light given out is of only a few discrete wavelengths, and each wavelength depends on the element, so that different elements give out different sets of wavelengths. Since these wavelengths are characteristic of the element, this type of emitted light can be used to identify the atoms producing it. This is the principle of *emission* spectrometry. All that one needs is to take a recording of the wavelengths emitted (usually this is in the form of a photographic plate where the position of an emission line on the plate depends on its wavelength). Fig. 8.19 shows spectrograms for the elements (a) sodium and (b) lithium. These can be used to confirm the presence of these elements in the spectrogram of Fig. 8.20, which is for a mixed light source.

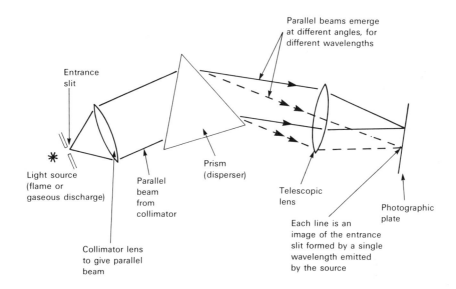

Fig. 8.20   Production of an emission spectrogram

The system used to produce an emission spectrogram is shown in Fig. 8.20. It consists of:

(a)   The light source — to make the individual atoms emit their characteristic wavelengths. They could be in a flame or in a gaseous discharge for example.

(b)   A focusing system (collimator), to produce a parallel beam of light.

(c)   A disperser of some type (prism or diffraction grating).

(d) A second focusing system (telescope) to bring the light down to a separate image of the slit for each wavelength of the light.

(e) A detector — eye, photographic plate or photocell.

The problem with this method of material identification is that it destroys the material, or means that the atoms must be able to be vaporised. This is neither always convenient nor always possible. There is a second method of identification called *absorption spectrometry*. Here the system works on the light absorbed by a material. The rule is that the wavelengths an element would emit at some temperature it will also absorb at the same temperature. Now the light source is an incandescent source giving out all visible wavelengths. A cell of the test material is placed in the beam of light from the source, and missing wavelengths are looked for on the photographic film (or some other detector system — a typical one is a chart recorder readout).

## EXERCISE 8

1) A wave train consists of 80 wavelengths of green light ($\lambda = 500$ nm). If the velocity of the wave is $3.00 \times 10^8$ m s$^{-1}$ for how long did the atom emit light?

2) A radio station sends out a ground wave, and a wave which is reflected from the ionosphere 90 km above the earth. What would be the wavelength of the signal to give a path difference of $40\lambda$ at 1000 km from the station on ground level?

3) At what angle is the second off-axis maximum ($n = 2$) for two coherent sources of waves ($\lambda = 600$ nm) if the source separation is 3 mm? It is assumed that the waves from the two sources come off parallel (see Fig. 8.12(b)).

4) Calculate the directions of reinforcement (constructive interference) for a grating of spacing $2.00 \times 10^{-6}$ m using light of wavelength $0.5\,\mu$m.

5) A grating has 6500 lines per cm. Calculate the angular separation of the wavelengths 589.6 nm and 546.1 nm respectively after transmission through at normal incidence, in the first-order spectrum.

6) Red light of wavelength 650 nm is incident normally in a diffraction grating of $6.000 \times 10^5$ lines per metre. At what angle is the first-order diffraction visible? How many orders of diffraction are visible? How many diffracted beams can be observed?

7)   A rectangular piece of glass 2 cm × 4 cm has 20000 evenly spaced lines ruled across its whole surface, parallel to the shorter side, to form a diffraction grating. Parallel rays of wavelength $5.5 \times 10^{-7}$ m fall normally on the grating. What is the highest order of the spectrum in the transmitted light.

# THE ELECTROMAGNETIC SPECTRUM

## ELECTROMAGNETIC WAVES

Examples of this type of wave are X-rays, ultra-violet light, visible light, infra-red, microwaves and radiowaves. Table 9.1 lists the properties of the whole range of electromagnetic waves. They have a wide range of wavelength and frequency. They have three common characteristics:

(a)  they consist of a varying electric effect (field) and a varying magnetic effect (field) which are in step (phase) but at right angles (see Fig. 9.1(a));

(b)  they are transverse waves (see Chapter 7);

(c)  they can travel through a vacuum and do so with a speed of $3.00 \times 10^8 \, \text{m s}^{-1}$ (300 million metres per second).

Fig. 9.1(a) depicts an electromagnetic wave. We can effectively ignore the magnetic effect and think of the wave as a normal transverse type (see Fig. 9.1(b)).

## MICROWAVES

Table 9.1 shows that wavelengths from about 1 mm to 30 cm are termed *microwaves*. They are produced most effectively by electronic methods in 'Klystron' tubes, where wavelength and power can be precisely controlled. The eye cannot detect microwaves, so that we use an electronic probe detector. Microwave sources for educational purposes have a wavelength of about 3 cm. Thus diffraction and interference experiments can be demonstrated on a large scale and measurements can be made using a centimetre rule. Some experiments which you can do are now described.*

*Teachers may also like to read the article 'Microwave Optics' by R Diamond in *School Science Review* (Nov. 1963) pp. 23–43.

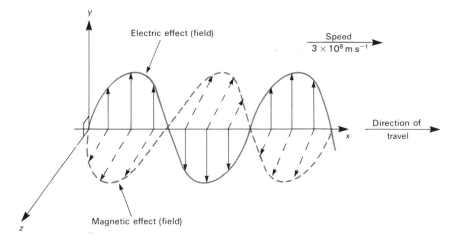

(a)  Relative orientation of electric and magnetic fields in an electromagnetic wave

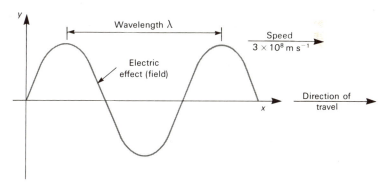

(b)  The wave shown as a simple transverse type

Fig. 9.1    An electromagnetic wave

## INTERFERENCE USING MICROWAVES

In Fig. 9.2 the source M has wavelength about 3 cm. Microwaves are absorbed and reflected by metal so that the two slits $S_1$ and $S_2$ act as small coherent sources of microwaves. There is a two-source interference pattern in the region beyond the slits which the probe T detects. (Note this set-up is the microwave equivalent of the two-source interference patterns described in Chapter 8.) Fig. 9.3 is a typical diagram of the interference pattern showing the nodal (destructive interference) and antinodal (constructive interference) lines. The zero-order anti-nodal line lies along the perpendicular bisector XX of PQ if M is arranged to be equidistant from P and Q. In practice it is convenient to arrange for this to be the case.

TABLE 9.1   Properties of Electromagnetic Waves

| Wavelength range | Name of radiation | Source | Method of detection | Main effects and uses |
|---|---|---|---|---|
| Less than $10^{-10}$ m | Gamma rays | Radioactive nucleus | Photographic film Ionisation chambers | Penetrate matter, so used in radiography and radiotherapy |
| Less than $10^{-9}$ m | X-rays | X-ray tube | | Diffracted by crystals, so used in crystallography |
| $10^{-9}$ m to $4 \times 10^{-7}$ m | Ultra-violet radiation | Gas discharge tube | Fluorescent material | Causes tanning of skin |
| | | Mercury vapour lamp | Photographic film | Causes fluorescence in visible region |
| | | | Photocell | Absorption and fluorescence spectra used in chemical analysis |
| $4 \times 10^{-7}$ m (violet) to $7.5 \times 10^{-7}$ m (red) | Visible light | Hot bodies (e.g. filament lamp) Gas discharge tube | Eye retina Photographic film Photocell | Gives sensation of vision Emission and absorption spectra used in chemical analysis |
| $7.5 \times 10^{-7}$ m to $1 \times 10^{-3}$ m | Infra-red radiation | Hot bodies | By heating effect (e.g. thermopile) | Heating effect used in infra-red therapy |
| | | | Special photographic film | Used in night time photography. Infra-red spectra give information on molecular structure |
| | | | Photoconductive cells Semiconductor devices | |
| $1 \times 10^{-3}$ m to $3 \times 10^{-1}$ m | Microwaves | 'Klystron' tube | Valve circuit Semiconductor devices | Demonstration of wave properties on a large scale Information on molecular structure |
| More than $3 \times 10^{-1}$ m | Radio waves | Radio aerials connected to special circuits | Tuned electrical circuits | Radio communication |

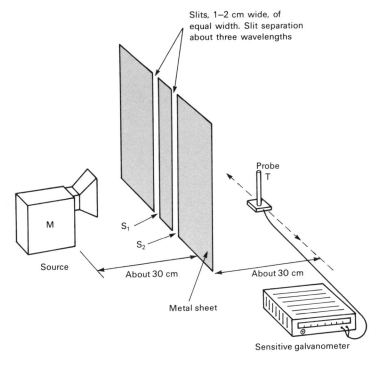

Slits, 1–2 cm wide, of
equal width. Slit separation
about three wavelengths

Probe
T

M

S₁

S₂

Source

About 30 cm

About 30 cm

Metal sheet

Sensitive galvanometer

**Fig. 9.2    An experimental set-up to observe interference using microwaves**

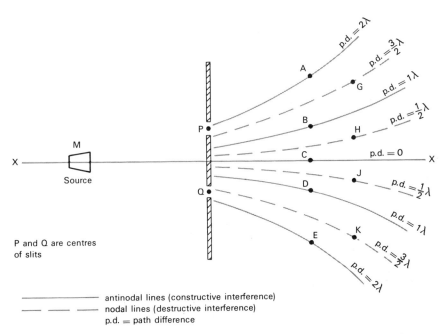

p.d. = 2λ

p.d. = 3/2 λ

p.d. = 1λ

p.d. = 1/2 λ

A

G

B

H

P

C    p.d. = 0

M

X — Source — X

D    J    p.d. = 1/2 λ

Q

p.d. = 1λ

E    K

p.d. = 3/2 λ

p.d. = 2λ

P and Q are centres
of slits

———————— antinodal lines (constructive interference)
— — — — nodal lines (destructive interference)
p.d. = path difference

**Fig. 9.3    A typical interference pattern using microwaves**

Referring to Fig. 9.3, suppose that positions of constructive interference are found at A, B, C, D, and E. We know that (see Chapter 8):

$$QA - PA = 2\lambda$$

$$QB - PB = \lambda$$

$$PD - QD = \lambda$$

$$PE - QE = 2\lambda$$

It is thus possible to find an average value for the wavelength $\lambda$ of the source M. We may also use the positions of destructive interference (in practice the nodes can often be more precisely located). In this case the relationships are:

$$QG - PG = \tfrac{3}{2}\lambda$$

$$QH - PH = \tfrac{1}{2}\lambda$$

$$PJ - QJ = \tfrac{1}{2}\lambda$$

$$PK - QK = \tfrac{3}{2}\lambda$$

### WORKED EXAMPLE 1

An experiment was set up using the apparatus shown in Fig. 9.2, in order to observe a two-source interference pattern using microwaves. The positions of constructive and destructive interference were located as shown in Fig. 9.4. Use this information to find an average value for the wavelength of the microwaves emitted by the source.

### Solution

The relevant distances may be found by scale drawing or by Pythagoras. Since the points are symmetrically situated above and below the line SC we shall use only those points above SC.

We find:

(a)     $QG = 40.0$ cm  and  $PG = 35.8$ cm

$\therefore$                  $QG - PG = \tfrac{3}{2}\lambda = 4.2$ cm

$\therefore$                      $\lambda = 2.8$ cm

(b)     $QB = 35.1$ cm  and  $PB = 32.1$ cm

$\therefore$                  $QB - PB = \lambda = 3.0$ cm

$\therefore$                      $\lambda = 3.0$ cm

(c)     $QH = 31.8$ cm  and  $PH = 30.2$ cm

$\therefore$                  $QH - PH = \dfrac{\lambda}{2} = 1.6$ cm

$\therefore$                      $\lambda = 3.2$ cm

The mean wavelength $\overline{\lambda}$ is the average of these three wavelength values. Thus

$$\lambda = (3.0 \pm 0.2)\,\text{cm}$$

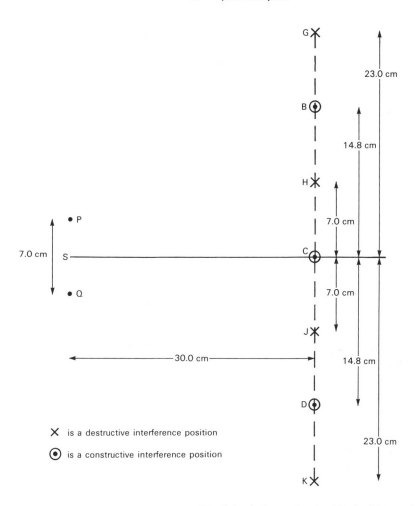

Fig. 9.4    Information for Worked Example 1

# DIFFRACTION USING MICROWAVES

Fig. 9.5 shows a simple set-up to demonstrate the diffraction of microwaves on passing through a single slit. The intensity contour of the diffraction pattern can be found using the detector probe T. The dependance of intensity contour on the slit size can be investigated. A typical diffraction pattern is shown in Fig. 9.6.

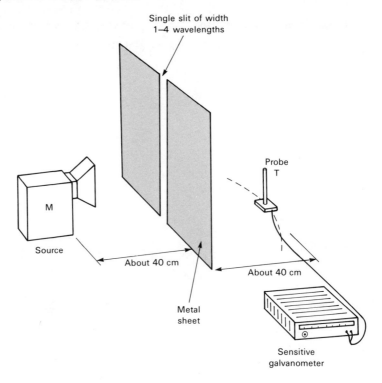

Fig. 9.5    Apparatus to observe single-slit diffraction using microwaves

A diffraction grating set-up for microwaves is shown in Fig. 9.7. The grating can be made from hacksaw blades, about 1 cm wide, placed equidistantly and about two wavelengths apart. (Alternatively, use strips of aluminium foil stuck on to a thin non-metallic sheet.*) The position and angle of the diffraction maxima can be located using the receiver R. Fig. 9.8 shows a typical pattern. If the source is a reasonable distance from the grating, we can assume that the microwaves arrive at the grating with plane wavefronts (see Chapter 8). Thus the relationship $d \sin \theta = n\lambda$ of equation [8.1] applies.

That is, referring to Fig. 9.8,

$$d \sin \theta_1 = 1\lambda$$
$$d \sin \theta_2 = 2\lambda$$

and so on for higher orders. From measurement of $d$ and $\theta$ values the mean wavelength can be determined.

*See the reference in the footnote on p. 124.

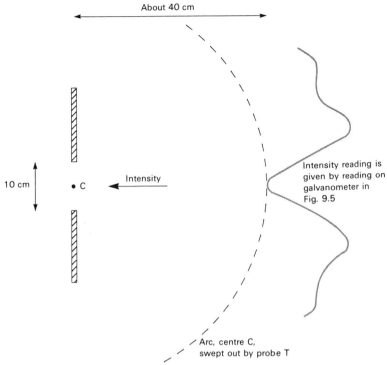

About 40 cm

10 cm

Intensity

• C

Intensity reading is given by reading on galvanometer in Fig. 9.5

Arc, centre C, swept out by probe T

Fig. 9.6 A single-slit diffraction pattern using microwaves of wavelength 3 cm

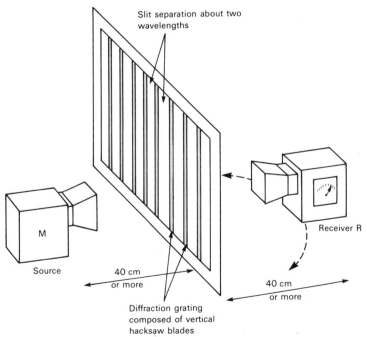

Slit separation about two wavelengths

M

Source

Receiver R

40 cm or more

40 cm or more

Diffraction grating composed of vertical hacksaw blades

Fig. 9.7 A diffraction grating set-up using microwaves

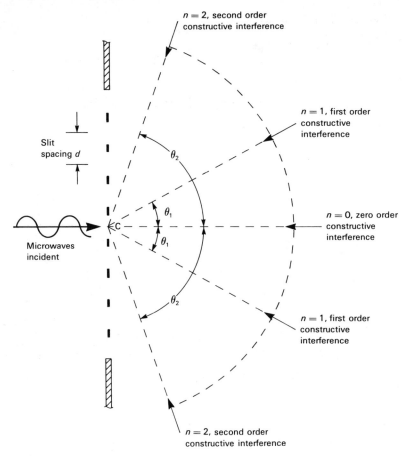

Fig. 9.8    Angular positions of diffraction maxima

The experiment requires some care if the correct positions of the diffraction maxima are to be identified. The arrangement of Fig. 9.7 is only a rough approximation to the proper diffraction grating set-up and thus our diffraction pattern has some spurious maxima present.

### WORKED EXAMPLE 2

An experiment was performed using apparatus similar to Fig. 9.7. The diffraction grating has a separation of 5.6 cm between the centres of the slits. Fig. 9.9 shows the positions of the zero- and first-order diffraction maxima. Use this information to find a value for the mean wavelength of the microwaves emitted by the source.

*Solution*

Rearranging equation [8.1] gives

$$\lambda = \frac{d \sin \theta}{n}$$

We have $n = 1$ and $d = 5.6$ cm. So:

(a)  for $\theta_1 = 34°$ then

$$\lambda = d\sin\theta_1 = 5.6\sin 34° = 3.13\,\text{cm}$$

(b)  for $\theta_1 = 31°$ then

$$\lambda = d\sin\theta_1 = 5.6\sin 31° = 2.88\,\text{cm}$$

∴ Mean wavelength $\overline{\lambda} = (3.0 \pm 0.1)$ cm.

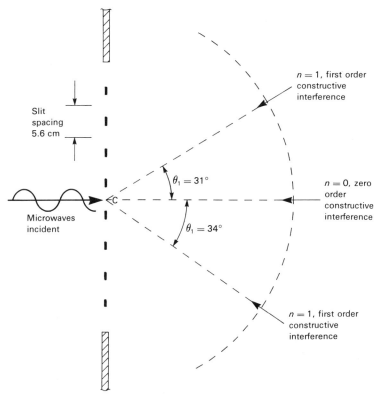

Fig. 9.9   Information for Worked Example 2

## POLARISATION USING MICROWAVES _____

Fig. 9.10 illustrates an experiment which shows that the microwaves from the source are plane-polarised (see Chapter 7). In Fig. 9.10(a) the intensity of microwaves received by the receiver R is hardly affected by the presence of the metal grid. This is because the metal rods are perpendicular to the direction of the varying electric field. If the grid is now rotated through $90°$, as in Fig. 9.10(b), then R receives hardly any microwave energy. This is because a rod more effectively 'stops' the microwaves when its length is parallel to the varying electric field.

Since this experiment shows that the microwaves from the source are plane-polarised, then it follows that microwaves are transverse waves.

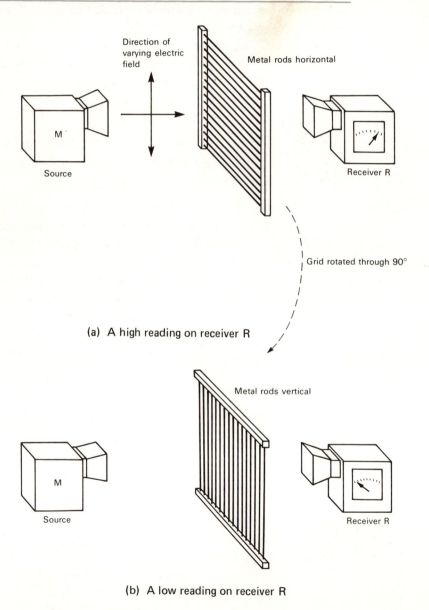

(a)  A high reading on receiver R

(b)  A low reading on receiver R

Fig. 9.10    Experiment to show that the microwaves from the source are plane-polarised

# INFRA-RED WAVES

Table 9.1 (page 126) shows that electromagnetic waves with wavelengths from about $0.75\,\mu$m to 1 mm are termed *infra-red* (or *IR*) *radiation*. This radiation is emitted by a hot body due to the energy given to the outermost electrons from the vibrations of its atoms and molecules. As the body gets hotter, the wavelengths emitted get shorter. An electric fire, or similarly hot body, radiates infra-red radiation which we can detect due to the rise in temperature which occurs when the radiant energy is absorbed by the skin. This warming effect is used in the treatment of injuries by infra-red lamps, since an increased skin temperature causes dilation of the blood vessels. Healing occurs more quickly due to increased blood flow.

The eye does not detect infra-red radiation (an electric fire is hot enough to give out *visible* red radiation which is why it glows visibly) so that special detectors must be used (see Table 9.1).

Fig. 9.11 shows how the presence of infra-red can be detected in the radiation emitted by a hot source. The prism disperses the radiation (see Chapter 8) so that wavelength increases in the direction blue to red and beyond. If the infra-red detector D is moved from the blue visible through to the red visible, a small reading is seen on the milliammeter. However, if the detector is now moved into the infra-red region, where wavelengths longer than red exist, then an increased meter reading is observed. This experiment indicates the existence of invisible radiation with a wavelength longer than red light. This is infra-red radiation. It is, like visible light, a transverse electromagnetic wave.

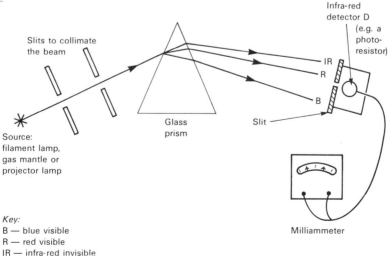

Key:
B — blue visible
R — red visible
IR — infra-red invisible

Fig. 9.11   Apparatus to show the presence of short wavelength infra-red radiation

The experiment can be extended by placing a filter in the beam which absorbs the visible but which allows the infra-red to pass through. Thus the region from blue to visible red gives an even smaller reading on the meter (mainly due to 'background' radiation). However, a high reading is still obtained in the infra-red region.

The above experiments deal with the shorter-wavelength infra-red radiation since the longer-wavelength radiation, say above 5 $\mu$m is absorbed by the glass of the prism. Fig. 9.12 shows a source of long-wavelength infra-red radiation. The 'Leslie' cube is a metal tank with its vertical faces painted matt black, shiny black, matt white and shiny white. When filled with hot water it emits mainly long-wavelength infra-red radiation which is detected by the thermopile P. The apparatus may be used to show that:

(a)    a matt black surface is the best emitter of infra-red radiation, and

(b)    a sheet of glass will cut off much of the emitted radiation.

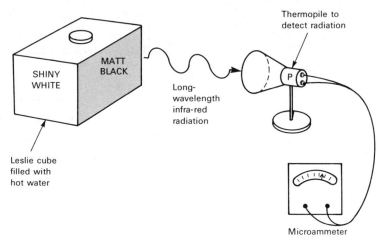

Fig. 9.12    Apparatus to show the properties of long-wavelength infra-red radiation

## ULTRA-VIOLET WAVES

Table 9.1 shows that electromagnetic waves with wavelengths from 0.4 $\mu$m down to about 0.001 $\mu$m are termed *ultra-violet* (or *UV*) *radiation*.

Ultra-violet waves are invisible to the naked eye. The longer-wavelength ultra-violet stimulates the production of vitamins in the skin and has beneficial effects. It is the component of sunlight which causes tanning. The shorter-wavelength ultra-violet is dangerous and can cause eye damage. Care must be taken to avoid looking directly at a source of ultra-violet radiation.

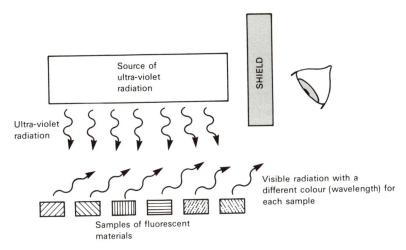

Fig. 9.13     The phenomenon of fluorescence

The presence of ultra-violet radiation can be detected by using materials which absorb ultra-violet and convert it to visible radiation. This is called fluorescence, and Fig. 9.13 shows how the phenomenon may be demonstrated. Many materials will fluoresce and washing-up powder, Vaseline® and antifreeze are common examples.

Fig. 9.14 is a schematic illustration of an apparatus which can be used to illustrate some of the properties of ultra-violet radiation. (Any lenses or prisms used should preferably be made of quartz since glass absorbs most ultra-violet radiation.) It is assumed that the dispersing mechanism produces

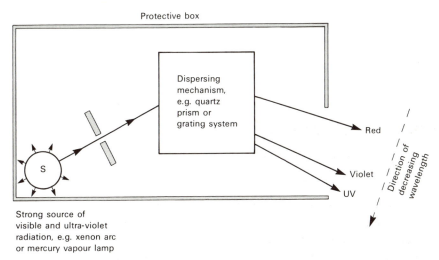

Fig. 9.14     Apparatus to demonstrate the properties of ultra-violet radiation

a spectrum as shown, so that the wavelength decreases in the direction shown. A sample of the fluorescent material placed in the invisible ultra-violet region beyond the violet will show the presence of radiation like visible light but with a shorter wavelength than violet light. In addition, a piece of photographic film will be blackened if placed in the ultra-violet region and a photoelectric cell will give a reading.

If specialist equipment is not available then a slide projector can be used as a source, a glass prism as a dispersing mechanism and photographic paper as a detector.*

## ORIGIN OF ULTRA-VIOLET RADIATION

A gas discharge tube is often used as a source of ultra-violet radiation. The electrical discharge excites the electrons in the atoms (see Chapter 14) and when these fall back to their normal state, their excess energy can be emitted as ultra-violet radiation. This is the same process as that by which visible radiation is emitted (see Chapter 8), but the energy jumps are higher for the ultra-violet region. This means that the individual bundles of wave energy (photons) of ultra-violet radiation have higher energy and hence shorter wavelength than visible radiation. Thus ultra-violet radiation has a lower wavelength than the violet light of the visible spectrum. Like visible light, it is a transverse electromagnetic wave.

## EXERCISE 9

1)   Refer to Fig. 9.15 which shows a double-slit interference set-up using microwaves, similar to that in Fig. 9.4. Assume that the source is equidistant from P and Q.

(a)   Calculate the mean wavelength of the source if $PQ = 8.0$ cm, $SC = 40.0$ cm, $CH = CJ = 8.4$ cm and $CB = CD = 16.0$ cm.

(b)   If the wavelength of the source is 3.0 cm, if $PQ = 9.0$ cm and $SC = 30.0$ cm, estimate by scale drawing the approximate values of $CH (= CJ)$ and $CB (= CD)$ you would expect.

2)   Refer to Fig. 9.16. Two identical microwave sources $M_1$ and $M_2$ of wavelength 3.0 cm face each other and are placed 60 cm apart. Describe what is observed if a probe is moved along the line joining $M_1$ and $M_2$ and around the midpoint position P.

*Teachers may refer to *Certificate Physical Science* Book 2 by A J Mee and A C E Jarvis (London, Heinemann, 1977 for further details.

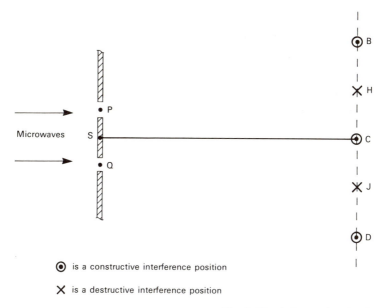

is a constructive interference position

is a destructive interference position

Fig. 9.15    Information for Question 1

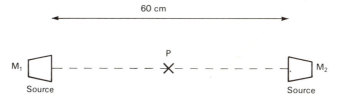

Fig. 9.16    Information for Question 2

3)    A beam of microwaves is incident normally on a (microwave) diffraction grating with slit spacing 6.0 cm. If a first-order diffraction maximum is observed at an angle of 28° to the forward direction, calculate the wavelength of the microwaves used.

4)    Calculate the angular separation of the two first-order diffraction maxima when a grating with slit spacing 5.6 cm has microwaves of wavelength 2.8 cm incident normally upon it.

5)    What kinds of electromagnetic waves would be emitted from a welder's arc?

6)    From your knowledge of the properties of short- and long-wavelength infra-red radiation, explain why the inside of a greenhouse can be hotter than its surroundings.

7)   What colour should central heating radiators be painted to maximise their radiating power?

8)   Some snakes have infra-red detectors, as well as eyes, in their heads. Why is this?

9)   Write down what precautions are necessary when using ultra-violet sources.

# 10

# SIMPLE HARMONIC MOTION

## OSCILLATIONS

An oscillating body is one which continuously retraces its motion. Simple harmonic motion (SHM) is a special type of oscillatory motion. A mass on the end of a spring and a pendulum bob on a string both perform SHM when disturbed by a small amount from their rest, or equilibrium, positions. The bodies oscillate because a restoring force acts to return the body to its equilibrium position. The motion is simple harmonic because the restoring force, provided by the spring or the weight of the bob, is proportional to the displacement of the mass or pendulum bob from its equilibrium position.

## CHARACTERISTICS OF SHM

Fig. 10.1(a) shows the instantaneous position of a particle P which performs SHM about its equilibrium position O. In one oscillation, or cycle, it traces out the path O to X to O to Y to O.

Fig. 10.1(b) is a plot of displacement versus time for the particle P. It is a sine curve (or a cosine curve, dependent upon where the origin of time is taken).

The periodic time $T$ is the time it takes the particle to make one complete oscillation.

The amplitude $r$ is the maximum distance of the particle from its equilibrium position.

In SHM $T$ is independent of $r$.

The velocity of the particle at any instant is given by the slope of the displacement–time curve. The velocity–time graph is shown in Fig. 10.1(c). The acceleration of the particle at any instant is given by the slope of the velocity–time curve. The acceleration–time graph is shown in Fig. 10.1(d).

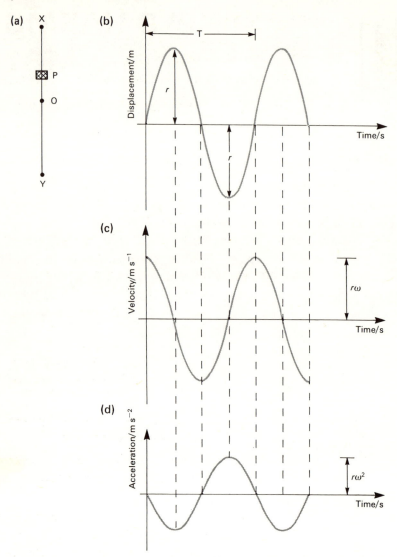

Fig. 10.1    (a) A particle P in SHM, (b) its displacement versus time, (c) its velocity versus time, (d) its acceleration versus time

Note the following points from Fig. 10.1(b), (c) and (d):

(a)    at maximum displacement ($\pm r$):

    (i) the velocity of the particle is zero since it is at the limits of its motion;

    (ii) the acceleration is a maximum, negative or positive, since the maximum restoring force acts at the limits of the motion;

(b)   at zero displacement:

(i) the particle has maximum velocity, positive or negative, as it passes through the centre position;

(ii) the acceleration is zero since no restoring force exists here.

Fig. 10.2 is a velocity–*displacement* graph obtained by combining Fig. 10.1(b) and (c).

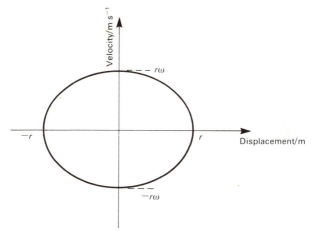

Fig. 10.2    Velocity versus displacement for SHM

Fig. 10.3 is an acceleration–*displacement* graph obtained by combining Fig. 10.1(b) and (d). It shows two important features of SHM:

(a)   The acceleration is positive when the displacement is negative and vice versa. This is because the restoring force acts to return the body to its equilibrium position.

(b)   The acceleration is proportional to the displacement from the equilibrium position. This is because the restoring force is proportional to the displacement — which means the system obeys Hooke's law. (A spring obeys Hooke's law and so will many other systems, if the displacement is small.)

The two features described above are characteristics which distinguish SHM from other types of oscillatory motion. Hence:

A particle is said to be oscillating in simple harmonic motion if it is moving about a point (O in Fig. 10.1(a)) in such a way that at any instant *its acceleration is directed toward the point and is proportional to the displacement from that point.*

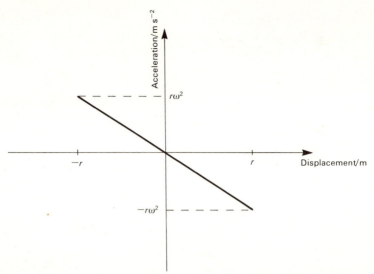

Fig. 10.3    Acceleration versus displacement for SHM

In mathematical terms the acceleration $a$ is related to the displacement $y$ by

$$a = -\text{Constant} \times y \qquad [10.1]$$

where the negative sign indicates that the acceleration is in the opposite direction to the displacement. Equation [10.1] is the mathematical form of Fig. 10.3.

### WORKED EXAMPLE 1

A particle performs SHM with an amplitude of 6.0 cm and a periodic time of 18 seconds. Table 10.1 shows time (column 2) and displacement (column 3) values over one complete oscillation and using these figures a displacement-time graph has been drawn as shown in Fig. 10.4. By drawing and measuring gradients at appropriate points, plot out the following graphs:

(a)    velocity-time and velocity-displacement;
(b)    acceleration-time and acceleration-displacement.

*Solution**

(a)    To determine the velocity profile we draw and measure gradients at the points marked 0, A, B, . . . , on Fig. 10.4. The values thus obtained are shown in Table 10.1, column 4. From columns 4 and 2 a velocity-time graph is constructed as shown in Fig. 10.5 (note the similarity to Fig. 10.1(c)). From columns 4 and 3 a velocity-displacement graph is constructed as shown in Fig. 10.6 (note the similarity to Fig. 10.2).

*Because of the symmetrical shape of the graphs it is necessary to make measurements of gradients over the first quarter cycle only, that is from O to E.

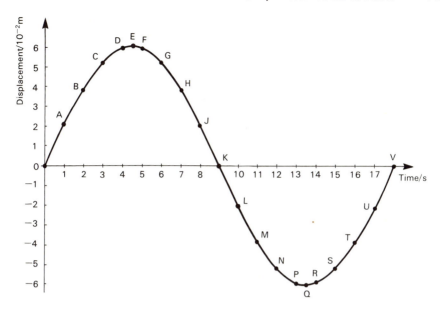

Fig. 10.4 Displacement versus time — information for Worked Example 1

(b) To determine the acceleration profile we draw and measure gradients at the appropriate points on the curve of Fig. 10.5. The acceleration values are shown in column 5 of Table 10.1. From columns 5 and 2 an acceleration-time graph is constructed as shown in Fig. 10.7 (note the similarity to Fig. 10.1(d). From columns 5 and 3 an acceleration-displacement graph is constructed as shown in Fig. 10.8 (note the similarity to Fig. 10.3).

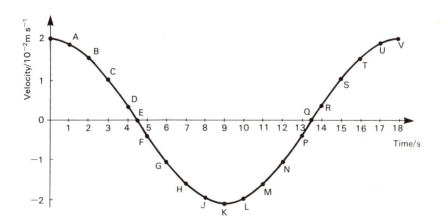

Fig. 10.5 Velocity versus time — solution for Worked Example 1

TABLE 10.1    Information and solution for Worked Example 1

| 1. Point (see Fig. 10.6) | 2. Time/s | 3. Displacement $10^{-2}$ m | 4. Velocity $10^{-2}$ m s$^{-1}$ | 5. Acceleration $10^{-2}$ m s$^{-2}$ |
|---|---|---|---|---|
| 0 | 0 | 0.00 | 2.09 | 0.00 |
| A | 1 | 2.05 | 1.96 | −0.25 |
| B | 2 | 3.86 | 1.60 | −0.47 |
| C | 3 | 5.20 | 1.05 | −0.63 |
| D | 4 | 5.91 | 0.36 | −0.72 |
| E | $4\frac{1}{2}$ | 6.00 | 0.00 | −0.73 |
| F | 5 | 5.91 | −0.36 | −0.72 |
| G | 6 | 5.20 | −1.05 | −0.63 |
| H | 7 | 3.86 | −1.60 | −0.47 |
| J | 8 | 2.05 | −1.96 | −0.25 |
| K | 9 | 0.00 | −2.09 | 0.00 |
| L | 10 | −2.05 | −1.96 | 0.25 |
| M | 11 | −3.86 | −1.60 | 0.47 |
| N | 12 | −5.20 | −1.05 | 0.63 |
| P | 13 | −5.91 | −0.36 | 0.72 |
| Q | $13\frac{1}{2}$ | −6.00 | 0.00 | 0.73 |
| R | 14 | −5.91 | 0.36 | 0.72 |
| S | 15 | −5.20 | 1.05 | 0.63 |
| T | 16 | −3.86 | 1.60 | 0.47 |
| U | 17 | −2.05 | 1.96 | 0.25 |
| V | 18 | 0.00 | 2.09 | 0.00 |

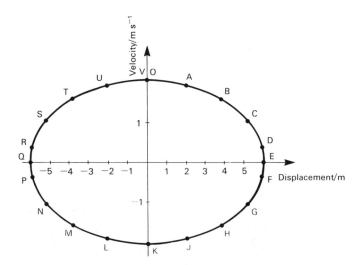

Fig. 10.6    Velocity versus displacement — solution for Worked Example 1

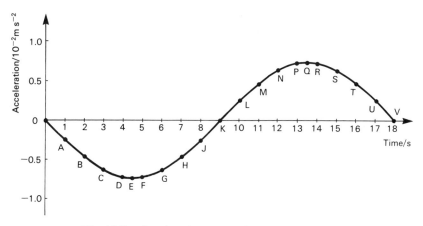

Fig. 10.7  Acceleration versus time — solution for Worked Example 1

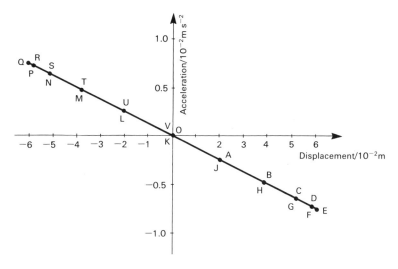

Fig. 10.8  Acceleration versus displacement — solution for Worked Example 1

## A SPRING–MASS SYSTEM

This is a simple system which performs SHM and it will be used to illustrate the characteristics of SHM discussed previously.

The spring in Fig. 10.9(a) obeys Hooke's law so that the relationship between force $F$, in newtons, and extension $e$, in metres, is

$$F = k \times e \qquad [10.2]$$

where $k$ is the force constant of the spring in newtons per metre.

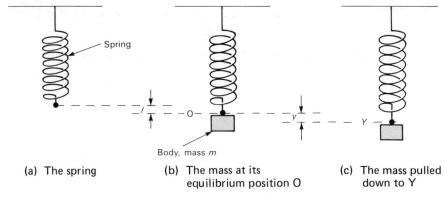

(a) The spring

(b) The mass at its
equilibrium position O

(c) The mass pulled
down to Y

Fig. 10.9    A spring–mass system

When a body of mass $m$, in kilograms, is attached to the base of the spring it extends by a distance $l$, where

$$mg = kl \qquad [10.3]$$

The body is now at its equilibrium position O (see Fig. 10.9(b)).

An *additional* force $ky$ is required to pull the body down a further distance $y$, to position Y in Fig. 10.9(c). When the body is released there will be an *instantaneous restoring force* $F = ky$ which accelerates the body toward its equilibrium position. The acceleration $a$ is found from $F = ma$ (see Chapter 1). That is:

$$ma = -ky \qquad [10.4]$$

or

$$a = -\left(\frac{k}{m}\right)y \qquad [10.5]$$

The negative sign indicates that the instantaneous force, and acceleration, is in the opposite direction to the displacement. The force thus acts to try to return the body to its equilibrium position. Also, since the restoring force is proportional to the displacement $y$, then so is the acceleration (see also equation [10.5]). Thus the mass oscillates with SHM.

Note that equation [10.5] is equivalent to equation [10.1], since $k$ and $m$ are constants for a given spring–mass system. On p. 154 we use equation [10.5] to show that the mass oscillates with periodic time $T$ given by

$$T = 2\pi \sqrt{\frac{m}{k}} \qquad [10.6]$$

*WORKED EXAMPLE 2*

(a)    A light helical steel spring of force constant $20\,\text{N m}^{-1}$ is hung vertically and a mass of $0.20\,\text{kg}$ hung from the lower end. Calculate the extension produced.

(b)   The mass is now pulled down a further distance of 2.0 cm and released. Calculate:

(i) the time period of subsequent oscillation,

(ii) the acceleration at the limits and at the centre of oscillation.

(Assume $g = 10\,\text{m s}^{-2}$.)

*Solution*

(a)   From equation [10.3]

$$mg = kl$$

$\therefore$    $$l = \frac{mg}{k} = \frac{0.20 \times 10}{20} = 0.10\,\text{m}$$

Thus when the mass is at its equilibrium position the spring is extended by 0.10 m.

(b)   (i) Periodic time $T$ is given by equation [10.6]:

$$T = 2\pi\sqrt{\frac{m}{k}} = 2\pi\sqrt{\frac{0.20}{20}} = 0.63\,\text{s}$$

Note that the periodic time is independent of amplitude.

(ii) The amplitude of the oscillation is 2.0 cm = 0.020 m.

At the limits of the oscillation $y = +0.020$ (displacement upwards) or $y = -0.020$ (displacement downwards). Using equation [10.5]:

A) for $y = +0.020$, acceleration $a$ is

$$a = \frac{-k}{m}y = \frac{-20}{0.20} \times 0.020 = -2.0\,\text{m s}^{-2}$$

B) for $y = -0.020$, acceleration $a$ is

$$a = \frac{-k}{m}y = \frac{-20}{0.20} \times -0.020 = +2.0\,\text{m s}^{-2}$$

Note that the acceleration is in the opposite direction to the displacement. At the centre of the oscillation, $y = 0$. Hence

$$a = \frac{-k}{m}y = 0$$

There is no acceleration at the centre since there is no restoring force when $y = 0$. However, since the body is moving it overshoots this position.

## PHASOR REPRESENTATION

Simple harmonic motion is related to uniform circular motion (see Chapter 2) and this relationship provides a useful way of representing SHM by what is called a *phasor.*

Fig. 10.10(a) shows a point moving with uniform angular speed $\omega$ about a circle of radius $r$ and centre O. This motion when projected on to diameter XY is the motion of the body P which, as shown in Fig. 10.10(b), performs SHM. Note that the radius $r$ of the circle equals the amplitude of the SHM. Thus the displacement–time variation in SHM is represented by the projection of the rotating radius. We call this *phasor representation* since, in addition to $r$, we need to know the phase angle to find the projected displacement.

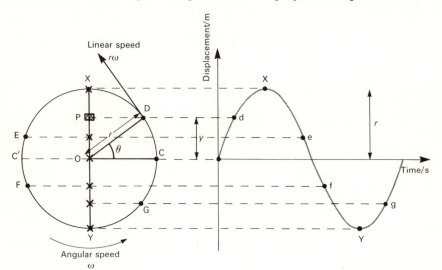

(a)  The point moving around        (b) The motion of P, or the motion of the
     the reference circle             point in (a) projected on to XY

Fig. 10.10    Phasor representation of SHM

Referring to Fig. 10.10(a), suppose at time zero the point is at C and at time $t$ the point is at D. Equation [2.2] gives the phase angle $\theta$, in radians, as

$$\theta = \omega t$$

The displacement $y$ of the body P is given by

$$\sin \theta = \frac{y}{r}$$

or                                    $$y = r \sin \theta \qquad\qquad\qquad [10.7(a)]$$

or                                    $$y = r \sin \omega t \qquad\qquad\qquad [10.7(b)]$$

When the point makes one revolution the body P performs one cycle of SHM. From equation [2.2]

$$\theta = \omega t$$

In one cycle the angle $\theta$ increases by $2\pi$ radians and time $t$ increases by the periodic time $T$ seconds. So equation [2.2] becomes

$$2\pi = \omega T$$

*or*
$$T = \frac{2\pi}{\omega}$$     [10.8]

Note that $\omega$ *is constant* for a given SHM and is called the *pulsatance*. It equals the angular speed of the point moving around the reference circle. Note however that SHM is a motion involving a *changing linear speed* along XY.

## WORKED EXAMPLE 3

A body oscillating in SHM moves in a vertical direction with an amplitude of 5.0 cm and a periodic time of 8.0 seconds. If it passes upwards through the equilibrium position at time zero, determine the displacement of the object at times 1 s, 2 s, 2.5 s, 5 s, 7 s, 8 s and 9 s later. In each case draw the phasor diagram to show the position of the rotating radius and the phase angle $\theta$ in degrees.

*Solution*

We are given $r = 5 \times 10^{-2}$ m and $T = 8$ s. The following relationships are needed:

from equation [10.7(b)]:     $y = r \sin \omega t$

from equation [10.8]:     $T = \dfrac{2\pi}{\omega}$

or     $\omega = \dfrac{2\pi}{T} = \dfrac{2\pi}{8} = \dfrac{\pi}{4}$

from equation [2.2]     $\theta = \omega t$

For $t = 1$ s

$$\theta = \omega t = \frac{\pi}{4}\text{rads} = 45°$$

$$y = r \sin \omega t = 5 \sin 45° = 3.54 \text{ cm}$$

The phasor is thus as shown in Fig. 10.11:

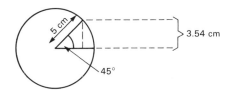

Fig. 10.11     Solution to Worked Example 3

The remaining solutions are shown in Table 10.2.

Note that for $t = 9\,\text{s}$ the body is into its second cycle.

### TABLE 10.2    Solution for Worked Example 3

| Time | Angle | | Displacement | Phasor |
|:---:|:---:|:---:|:---:|:---:|
| $t/s$ | $\theta/\text{rad}$ | $\theta/\text{degree}$ | $y/\text{cm}$ | diagram |
| 1 | $\dfrac{\pi}{4}$ | 45 | 3.54 | 45° |
| 2 | $\dfrac{\pi}{2}$ | 90 | 5.00 | 90° |
| 2.5 | $\dfrac{5\pi}{8}$ | $112\frac{1}{2}$ | 4.62 | 112½° |
| 5 | $\dfrac{5\pi}{4}$ | 225 | $-3.54$ | 225° |
| 7 | $\dfrac{7\pi}{4}$ | 315 | $-3.54$ | 315° |
| 8 | $2\pi\,(0)$ | 360 (0) | 0 | |
| 9 | $\dfrac{9\pi}{4}\left(=\dfrac{\pi}{4}\right)$ | 405 (= 45) | 3.54 | 45° |

## ACCELERATION AND VELOCITY VALUES

Instantaneous acceleration and velocity values may be obtained from Fig. 10.10(a). In Chapter 2 we noted that the point in Fig. 10.10(a) moves with uniform linear speed $r\omega$ and has acceleration $r\omega^2$ toward the centre. The speed $v$ and acceleration $a$ of body P are the components of $r\omega$ and $\overline{r\omega^2}$ along the vertical diameter. These are:

(a)    *at the centre of the motion*

|  |  |  |
|---|---|---|
| P moving up | $v = +r\omega$ | [10.9(a)] |
| P moving down | $v = -r\omega$ | [10.9(b)] |

and P moving up or down    $a = 0$

The acceleration is zero here since $r\omega^2$ has no vertical component at point C or C' (see Fig. 10.10(a)).

(b)  *at the limits of the motion*

| | | |
|---|---|---|
| P at X | $a = -r\omega^2$ | [10.10(a)] |
| P at Y | $a = +r\omega^2$ | [10.10(b)] |
| P at X and Y | $v = 0$ | |

The velocity is zero since $r\omega$ has no vertical component when the point is at X or Y. Note that the acceleration is downwards negative) when the displacement is upwards (positive) and vice versa.

In general, at some phase angle $\theta$,

$$v = r\omega \cos \theta$$

$$a = -r\omega^2 \sin \theta$$

and since $y = r \sin \theta$, we can show

$$v = r\omega\sqrt{r^2 - y^2} \qquad [10.11]$$

and $$a = -\omega^2 y \qquad [10.12]$$

Note that the equation [10.11] is the mathematical form of Fig. 10.6 and equation [10.12] the form of Fig. 10.8.

## WORKED EXAMPLE 4

A light helical steel spring of force constant $30 \text{ N m}^{-1}$ is suspended vertically and a mass of 0.20 kg added to the lower end. The mass is pulled down a further distance of 2.0 cm below the equilibrium position and released. Calculate:

(a)  the acceleration at the limits and at the centre of the oscillation,

(b)  the velocity at the limits and at the centre of the oscillation.

*Solution*

We must calculate $\omega$.

From equation [10.8]       $T = \dfrac{2\pi}{\omega}$   or   $\omega = \dfrac{2\pi}{T}$

From equation [10.6]       $T = 2\pi\sqrt{\dfrac{m}{k}}$

where $m = 0.20 \text{ kg}$ and $k = 30 \text{ N m}^{-1}$

$$\therefore \quad \omega = \frac{2\pi}{T} = \frac{2\pi}{2\pi\sqrt{\dfrac{m}{k}}} = \sqrt{\frac{k}{m}} = \sqrt{\frac{30}{0.20}}$$

$$= 12.25 \text{ rad s}^{-1}.$$

(a)  At the limits of motion, equations [10.10] give

$$a = \mp r\omega^2$$

where $r = 0.020$ m and $\omega = 12.25$ rad s$^{-1}$. Thus

at top of motion,        $a = -0.020 \times (12.25)^2$

$$= -3.0 \text{ m s}^{-2}$$

at bottom of motion,     $a = +0.020 \times (12.25)^2$

$$= +3.0 \text{ m s}^{-2}$$

At the centre of motion $a = 0$.

(b)  At the limits of motion $v = 0$.

At the centre of motion, equations [10.9] give

$$v = \pm r\omega$$

Hence:

when moving up,        $v = +0.020 \times 12.25$

$$= 0.245 \text{ m s}^{-1}$$

when moving down,      $v = -0.020 \times 12.25$

$$= -0.245 \text{ m s}^{-1}$$

## PERIODIC TIME AND FREQUENCY OF A SPRING—MASS SYSTEM: PROOF OF RELATIONSHIPS

The periodic time of SHM is from equation [10.8] given by $T = \dfrac{2\pi}{\omega}$. For the spring–mass system of Fig. 10.9 we compare equations [10.5] and [10.12] to identify $\omega^2 = \dfrac{k}{m}$. Thus the periodic time of the spring-mass system is given by

$$T = \frac{2\pi}{\omega} = \frac{2\pi}{\sqrt{\dfrac{k}{m}}}$$

or
$$T = 2\pi \sqrt{\frac{m}{k}}$$

as stated in equation [10.6].

The frequency of oscillation $f$ (the number of oscillations in 1 second) is related to $T$ by equation [7.1] which gives

$$f = \frac{1}{T}$$

or
$$f = \frac{1}{2\pi \sqrt{\frac{m}{k}}}$$

$$f = \frac{1}{2\pi} \sqrt{\frac{k}{m}} \qquad [10.13]$$

This is the frequency with which the spring–mass system will vibrate if disturbed and then allowed to oscillate in the absence of external forces. This is called the *natural frequency* and will be denoted by $f_N$ from now on.

# FORCED VIBRATIONS

We can make a system vibrate at a frequency not equal to its natural frequency — we can *force* it to vibrate at the same frequency as an externally applied force.

Fig. 10.12 shows how this can be simply demonstrated. The externally applied force comes from the hand which is moved up and down at the desired disturbing frequency or forcing frequency $f$ and with amplitude $D$. If $D$ is kept constant and the *frequency* of the disturbance altered, then we find that the amplitude $r$ of the mass varies.

The ratio $\dfrac{r}{D}$ is called the *transmissibility*.

Fig. 10.13 shows the relationship between transmissibility and forcing frequency for a system of natural frequency $f_N$. Note the features shown in Fig. 10.13:

(a)   For $f$ much less than $f_N$: $r = D$.
Moving the hand up and down slowly results in a displacement of the whole spring–mass system by the same amount.

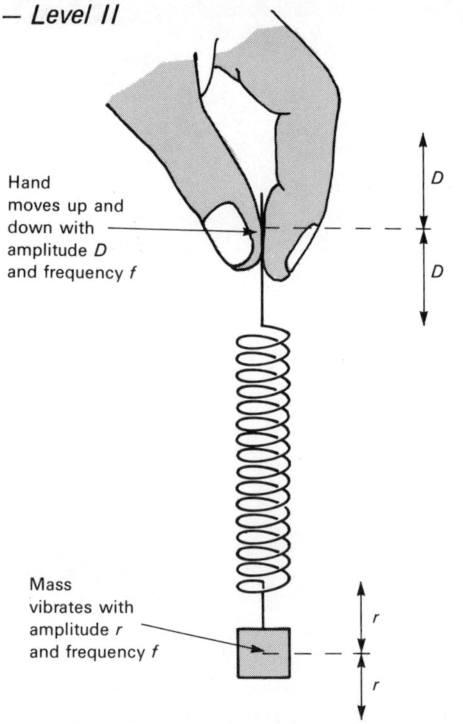

Hand
moves up and
down with
amplitude *D*
and frequency *f*

*D*

*D*

Mass
vibrates with
amplitude *r*
and frequency *f*

*r*

*r*

Fig. 10.12    Forcing a spring–mass system to vibrate with a frequency *f* determined
by the motion of the hand

(b)    For *f* much larger than $f_N$:  $r = 0$.
The disturber vibrates so fast that the inertia of the spring–mass
system makes it incapable of moving at all.

(c)    For *f* equal or nearly equal to $f_N$: *r* is greater than *D*.
This is because it is easy to transfer energy from the driving system (the
hand) to the driven system (spring–mass) when *f* is close to $f_N$.

## RESONANCE

Fig. 10.13 shows that *r* is a maximum and is much greater than *D* when
$f = f_N$. This situation is called *resonance*.

Resonance occurs when an oscillator has applied to it a periodically varying
force with frequency equal to the natural frequency of the oscillator.

## DAMPING

So far, the effect of external forces which dissipate energy during the oscillation have been ignored. Clearly these exist — for example, the oscillations
of a mass on a spring gradually die away. This is because the energy of the

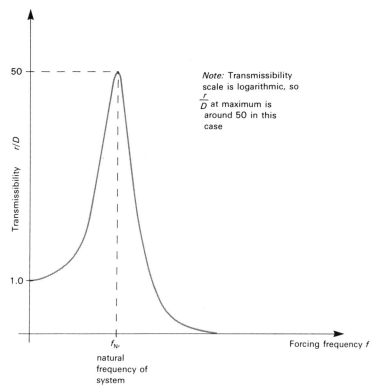

Fig. 10.13   Transmissibility versus forcing frequency for a system with natural frequency $f_N$

oscillatory motion is transferred to, for example, movement of air molecules as a result of air friction. The oscillations are referred to as 'damped'. For the case of air friction the amplitude of motion reduces slowly so it is termed 'light' damping. Immersion of the mass in heavy oil would cause the oscillations to reduce in amplitude very quickly. This is termed 'heavy' damping.

The effects of damping are as follows:

(a)   The amplitude of the oscillatory motion decreases with time.

(b)   The driving force frequency $f$ at which resonance occurs becomes progressively less than the natural frequency $f_N$ of the system as damping increases. However, with light damping the difference between $f$ and $f_N$ is negligible. (Note that Fig. 10.13 refers to light damping.)

(c)   The amplitude at resonance becomes less as damping increases.

As a result of (b) the statement about resonance becomes:

Under conditions of light damping, resonance occurs as a result of the applied frequency being equal to the natural frequency of the system.

# RESONANCE DAMAGE

At resonance the amplitude of the driven system can be very large. This can cause damage so we must take care.

Practical systems are much more complicated than a mass on a spring. A machine, since it has a number of parts, will have a number of natural frequencies. If the machine had a motor then when the motor started up the various parts would vibrate or rattle when the rotational frequency of the motor matched the natural frequency of the part. This can be noisy and irritating. It could cause damage and be dangerous if vibration resulted in failure or breakage.

# ANTIVIBRATION MOUNTINGS

Sensitive machinery must be protected from vibration caused by other machines. This can be done by the use of antivibration mountings (see Fig. 10.14). The mountings are chosen so that their transmissibility at the frequency of the disturbing machine is very small. Fig. 10.13 tells us we must choose $f_N$ much less than the disturbing frequency. In practice this means that the force constant or 'stiffness' $k$ of the mount in Fig. 10.14 must be low. The mount must thus be flexible.

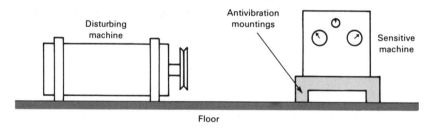

Fig. 10.14   Use of antivibration mountings

*WORKED EXAMPLE 5*

Fig. 10.15 shows a concrete block of mass 500 kg which is supported at its four corners by mechanisms which act like springs, each with force constant 500 N m⁻¹, once the block is in position. The block and springs form an antivibration mounting on to which sensitive instruments are placed. Calculate the natural frequency of vibration of the block–spring system and hence explain:

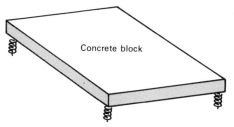

Fig. 10.15  An antivibration mounting — information for Worked Example 5

(a)  why vibrations from a 50 Hz motor will not affect instruments placed on the block,

(b)  why an operator must not stamp near the platform with a frequency of about 1 per second.

*Solution*

From equation [10.13]

$$f_N = \frac{1}{2\pi}\sqrt{\frac{k}{m}}$$

where $m = 500$ kg and $k = 500 \times 4 = 2000$ N m$^{-1}$ (since there are four springs)

$$\therefore \quad f_N = \frac{1}{2\pi}\sqrt{\frac{2000}{500}} = \frac{1}{2\pi}\sqrt{4} = \frac{1}{\pi}$$

$$= 0.318 \text{ Hz}$$

(a)  Since $f_N$ is much less than 50 Hz, the transmissibility at 50 Hz will be very small.

(b)  A periodic disturbance of about 1 Hz is close to resonance and the transmissibility could be large enough to cause appreciable movement of the block.

# ABSORPTION SPECTRA

Chemists require information about the nature and strength of molecular bonds and this can be gained as follows. Fig. 10.16 shows a simple model of a hydrogen chloride molecule. The complex electronic bond has been replaced by a simple spring of force constant $k$. (A discussion of why we can do this is given at the end of this chapter.) The chlorine is much more massive than the hydrogen so that when the molecule is disturbed only the hydrogen atom vibrates appreciably. It vibrates at the natural frequency $f_N$ which is characteristic of hydrogen chloride. Equation [10.13] gives

$$f_N = \frac{1}{2\pi}\sqrt{\frac{k}{m}}$$

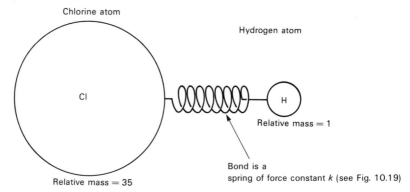

Fig. 10.16    A model of a hydrogen chloride molecule

where $m$ refers to the mass of the hydrogen atom and $k$ is the stiffness or *force constant* of the HCl bond.

Values for $f_N$ can be found from absorption spectra (see Chapter 8). The disturbing frequency arrives at the HCl molecule in the form of an electromagnetic wave of incident intensity $I_0$ as shown in Fig. 10.17. The intensity $I$ of the transmitted beam is measured and the ratio $I/I_0$, called the *absorption A* is calculated. This is done for a range of values of incoming frequency $f$. Fig. 10.18 shows the variation of $A$ with $f$. The dip in the graph occurs when $f = f_N$, since at resonance there is a large intake of energy from the incident beam. Since $f_N = \dfrac{1}{2\pi}\sqrt{\dfrac{k}{m}}$ and $m$ is known, we can find $k$, the force constant of the bond.

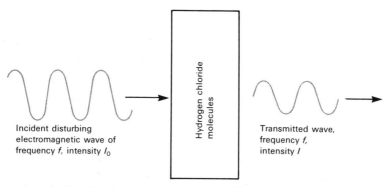

Fig. 10.17    An electromagnetic wave incident on a hydrogen chloride specimen

### WORKED EXAMPLE 6

The resonance absorption frequency for hydrogen chloride is $9.0 \times 10^{13}$ Hz. By considering the hydrogen chloride molecule as two masses attached to opposite ends of a

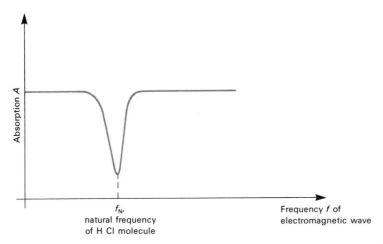

Fig. 10.18    Absorption versus incident frequency of electromagnetic wave

simple spring, deduce the force constant of the HCl bond. The mass of the hydrogen atom is $1.66 \times 10^{-27}$ kg and that chlorine atom is roughly 35 times more massive, and hence only the vibration of the hydrogen atom need be considered.

*Solution*

From equation [10.13]

$$f_N = \frac{1}{2\pi} \sqrt{\frac{k}{m}},$$

where $f_N = 9.0 \times 10^{13}$ and $m = 1.66 \times 10^{-27}$. Rearranging [10.13] gives

$$k = 4\pi^2 f_N^2 \times m = 4\pi^2 \times (9 \times 10^{13})^2 \times 1.66 \times 10^{-27}$$

$$= 531 \, N \, m^{-1}$$

Table 10.3 lists values of absorption frequency for a range of hydrogen halides. Our simple picture of the molecules gives the force constants in column 3 (these you can check using $m = 1.66 \times 10^{-27}$ kg). A more complex treatment gives the values shown in column 4. Note the good agreement.

TABLE 10.3    Absorption frequency and force for hydrogen halides

| 1. Hydrogen halide | 2. Absorption frequency/Hz | Deduced force constant/N m$^{-1}$ | |
| --- | --- | --- | --- |
| | | 3. Simple model | 4. Complex model |
| HF | $12.4 \times 10^{13}$ | 1008 | 966 |
| HCl | $9.0 \times 10^{13}$ | 531 | 516 |
| HBr | $7.9 \times 10^{13}$ | 409 | 412 |
| HI | $6.9 \times 10^{13}$ | 312 | 314 |

# A SPRING–MASS MODEL FOR THE HCI MOLECULE _____

We now discuss why we can represent the HCl molecule by H and Cl joined by a bond which has the characteristics of a simple spring.

The force separation curve between typical atoms is shown in Fig. 10.19(a). The relatively massive chlorine atom is at O and the much lighter hydrogen atom vibrates about the equilibrium separation position at C, with amplitude $r$ as shown. In Fig. 10.19(b) the line ACB represents the force $F$ between H

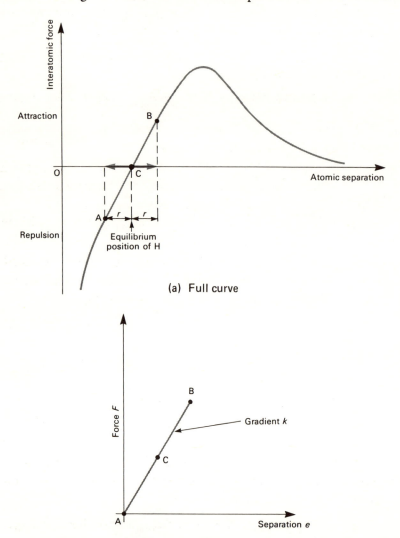

(a)  Full curve

(b)  Force versus separation from position A

Fig. 10.19    Force between hydrogen and chlorine atoms versus separation of atoms

and Cl as a function of separation $e$ from position A. Since ACB is a small distance, it can be treated as a straight line, then $F = ke$, where $k$ is the gradient of ACB. This is the same equation as a spring so we can *mathematically* replace the bond between H and Cl by a simple spring-type bond (provided the amplitude of the motion is not too large).

**EXERCISE 10** _____

1) Define simple harmonic motion. Hence explain why any system which obeys Hooke's law will vibrate in SHM when displaced.

2) In the following table, fill in the missing spaces for the magnitude of the quantities shown:

|     | *Displacement* | *Velocity* | *Acceleration* |
| --- | --- | --- | --- |
| (a) | ? | Maximum | ? |
| (b) | ? | ? | Maximum |

3) (a) When a helical spring, itself of negligible mass, has a mass of 0.30 kg applied to its lower end it extends by 0.080 m. Calculate:
   (i) the force constant of the spring;
   (ii) the time period and frequency of vertical oscillations of the mass.

   (b) Find the magnitude of the mass which, when suspended from the same spring, has a frequency of oscillation of 1.0 Hz. Assume acceleration due to gravity $g = 10 \, \mathrm{m \, s^{-2}}$.

4) A particle is vibrating with simple harmonic motion in a vertical direction. It has an amplitude of 12.0 cm and a periodic time of 5.00 s. Assume at $t = 0$ the displacement is zero and it is moving upwards. Draw phasor diagrams to represent the motion at one second intervals over a cycle and hence calculate the displacement at these times.

5) A body oscillates in SHM with a periodic time of 16 s and an amplitude of 4 cm. Assuming that it starts from zero displacement at time zero, draw phasor diagrams at one second intervals over one complete oscillation. Use this information to calculate the displacement of the particle at these times, and hence plot out the displacement–time curve.

6) The mains electrical supply in Great Britain has a voltage variation with time which is the same form as simple harmonic motion. It has a frequency of 50 Hz and a voltage amplitude of 339 V.
(a) Calculate the periodic time.

(b)    Use a phasor method similar to that used in Question 5 to calculate the voltage variation with time over a single cycle, and hence plot out the voltage time curve.

7)    A body oscillates in SHM with an amplitude of 3.0 cm and a frequency of 5.0 Hz. Calculate:

(a)    the periodic time;

(b)    the acceleration at the extremities and the centre of the oscillation;

(c)    the velocity at the extremities and at the centre of the oscillation.

8)    The 0.20 kg mass shown in Fig. 10.20 is tethered by two identical springs of force constant $5.0 \, \text{N m}^{-1}$. The mass is now displaced by 3.0 cm to the right from its equilibrium position and released. Calculate:

(a)    the time period and frequency of subsequent oscillations;

(b)    the acceleration at the extremities and at the centre of the oscillation;

(c)    the velocity at the extremities and at the centre of the oscillation.

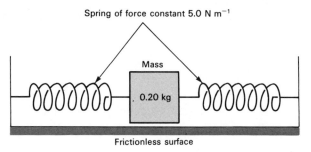

Fig. 10.20    Information for Question 8

9)    The piston in a car engine moves in approximately simple harmonic motion. If the mass of the piston is 0.80 kg and if its amplitude is 6.0 cm, calculate the maximum force experienced when the crankshaft rotates at 6000 r.p.m. (revolutions per minute).

10)    A body of mass 0.30 kg has a maximum force of 0.12 N acting on it when it oscillates in SHM with an amplitude of 2.0 cm. Calculate the periodic time of oscillation.

11)    The metal guard plate on an AC mains motor has a natural frequency of 50 Hz. What could be the problem for anybody working nearby?

12)    (a)    Define *resonance*. Explain why mechanical damage is likely to occur as a result of resonance.

(b) Describe some examples which you have heard about, or experienced, where resonance has been a nuisance.

13) (a) In the absorption spectra of an organic liquid there is a dip in the transmittance at $1.20 \times 10^{14}$ Hz. It is thought that this is due to oscillations of a hydrogen atom coupled to a more massive molecule. Deduce the force constant for the hydrogen bond involved. Assume that the mass of a hydrogen atom is $1.66 \times 10^{-27}$ kg.

(b) In the above experiment there is also a dip in the transmittance at $2.40 \times 10^{14}$ Hz. Suggest an explanation for this.

# ELECTROMAGNETIC FORCES AND THE MOVING-COIL INSTRUMENT

## MAGNETIC FIELDS

### THE CONCEPT OF A MAGNETIC FIELD

It is well known that the 'needle' of a magnetic compass will be deflected if a bar magnet is brought near to it. In some way the magnet exerts a force on the needle, and this force takes place without any physical contact between the magnet and the needle.

To explain this phenomenon we say that the bar magnet produces a magnetic field in the space around it, and it is this field that causes the needle to deflect.

To represent magnetic fields in diagrams we use lines, with an arrow for indicating direction. This is similar to using lines to represent rays of light. The magnetic field produced by a bar magnet is shown in Fig. 11.1. The direction of the field is the direction in which a magnetic compass needle points.

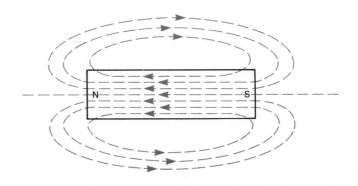

Fig. 11.1   The magnetic field of a bar magnet

It is sometimes convenient to identify the ends of a magnet as N (for North) and S (for South). If a bar magnet was suspended about its mid-point, the N end would point to the geographical North. It is thus inferred that:

(a)   Magnetic field lines leave the N-end and enter the S-end of a bar magnet.

(b)   The magnetic field lines form continuous loops which pass through the magnet.

Other magnetic field patterns are shown in Fig. 11.2. It should be remembered that the magnetic field lines do form continuous loops, although this is not always shown.

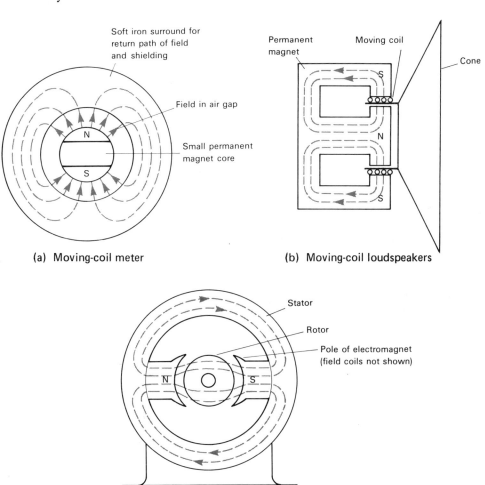

(a)  Moving-coil meter          (b)  Moving-coil loudspeakers

(c)  Two-pole d.c. motor

Fig. 11.2   Magnetic field patterns

## FIELDS FROM CURRENT

A magnetic field can also be produced by an electric current. If, for example, an electric current is passed through a cylindrical coil about the size of a bar magnet (Fig. 11.3), the pattern of the magnetic field produced is very similar.

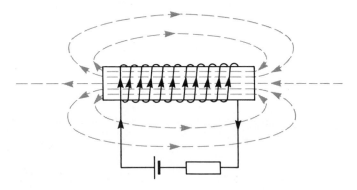

Fig. 11.3    The magnetic field produced by a current in a coil wound on a cardboard tube (solenoid)

The pattern of such fields can be found experimentally by using a magnetic compass or iron filings. If this is done for a straight wire or conductor carrying a current, it will be found that the magnetic field forms a series of concentric circles, as shown in Fig. 11.4.

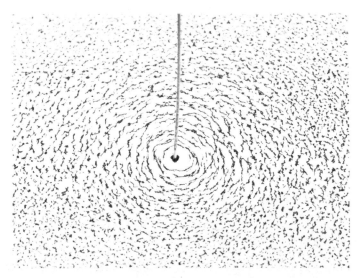

(a)  Iron filings show circular pattern of field due to current in wire

Fig. 11.4 (continued opposite)

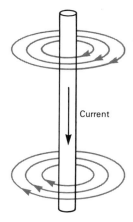

(b) Concentric magnetic field due to current in a straight conductor

Fig. 11.4

An interesting pattern is the magnetic field produced by a coil wound on a circular tube — known as a *toroidal coil* or *toroid*. This is shown in Fig. 11.5, and it should be noted that the field is entirely within the coil, and is uniform as shown by the magnetic field lines being equally spaced.

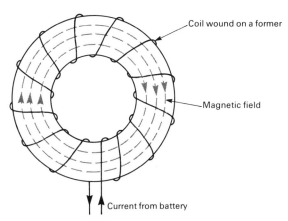

Fig. 11.5    The magnetic field produced by a toroid

## RULE FOR FINDING THE DIRECTION OF THE MAGNETIC FIELD

For a straight conductor the direction of the magnetic field produced by a current in it can be found easily as follows.

Imagine you are holding the wire in your right hand as shown in Fig. 11.6. Point your thumb in the direction of the current, your fingers will curl around in the direction of the magnetic field.

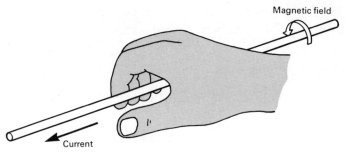

Fig. 11.6    Finding the direction of magnetic field due to a straight conductor

This can be adapted for a coil, as shown in Fig. 11.7. Grasp the coil so that this time your fingers point in the direction of the current in each turn of the coil, then your thumb will point in the direction of the magnetic field.

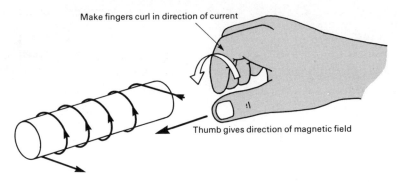

Fig. 11.7    Finding the direction of the magnetic field in a coil

An adaption of this rule is the *right-hand corkscrew rule*. If a corkscrew is pointed in the direction of the current in a wire as shown in Fig. 11.8, then the direction in which a corkscrew is normally turned (clockwise) will give the direction of the magnetic field. It is left for you to work out how this rule can be adapted for a coil.

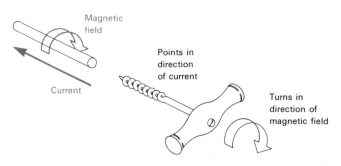

Fig. 11.8    The 'corkscrew' rule

An important point to note is that if the current direction is reversed, then the direction of the magnetic field is reversed.

## MAGNETIC FLUX (Φ) AND FLUX DENSITY (*B*)

Although the magnetic field around a magnet is not visible its effects are obvious. They are that in this region:

(a)    Another magnet experiences a force.

(b)    A wire carrying a current would also experience a force.

(c)    A moving search coil could generate an e.m.f. (see Chapter 12).

The field lines constitute a *magnetic flux*. The symbol for magnetic flux is Φ, and it is measured in webers (Wb). We draw field lines to represent a magnetic flux, and to distinguish between a region of strong field, and a region of weaker field, we specify the number of field lines per unit area or the *flux density B*. Then (see Fig. 11.9):

$$B = \frac{\Phi}{A} \qquad\qquad [11.1]$$

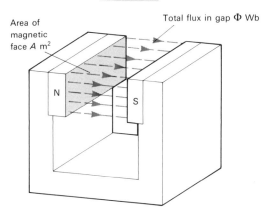

Fig. 11.9   Flux density $B = \Phi/A$

The unit of magnetic flux density is the tesla (T).

These units (Wb) and (T) are derived units from experiments described in Chapter 12.

### WORKED EXAMPLE 1

A student estimates that the magnetic flux in an iron toroid of the type shown in Fig. 11.5 is $2 \times 10^{-3}$ Wb. If the cross-sectional area of the toroid is $2 \text{ cm}^2$, is the estimated value sensible?

### Solution

We test this by finding the value of the magnetic flux density to see if it falls within the possible range for iron, which is not likely to be greater than 2 T. We have, from equation [11.1],

$$\text{Flux density } B = \frac{\text{Flux}}{\text{Area } (A)} = \frac{2 \times 10^{-3}\,\text{Wb}}{2 \times 10^{-4}\,\text{m}^2} = 10\,\text{T}$$

This is not reasonable; it is far too great.

### WORKED EXAMPLE 2

A current of 0.5 A produces a flux of $1 \times 10^{-6}$ Wb in a toroidal coil which is air-cored, i.e. there is no iron present. If in such a case the flux produced is proportional to current, what current would be required for a flux density of $1 \times 10^{-2}$ T? The cross-sectional area of the coil is 2 cm$^2$.

### Solution

For a flux density of $1 \times 10^{-2}$ T, the total flux will be given by

$$\Phi = BA \quad \text{(from equation [11.1])}$$
$$= (1 \times 10^{-2}) \times (2 \times 10^{-4})\,\text{Wb}$$
$$= 2 \times 10^{-6}\,\text{Wb}$$

This is twice the flux produced by 0.5 A, so the required current is 1.0 A.

## INTERACTION OF MAGNETIC FIELDS

The needle of a magnetic compass is itself a small magnet. As such it produces a magnetic field. This field interacts with the Earth's magnetic field, and the result is that the needle points North. Similarly the needle can be made to deflect by having a bar magnet brought near it. These are two examples of magnetic fields interacting.

When two magnetic fields interact, the individual magnetic fields are modified to produce one resultant pattern. It is well known that if the S-ends of two bar magnets are brought near each other, there is a repulsion force between them. If an N-end of one and the S-end of another are brought near, they attract each other. The two situations are shown in Fig. 11.10.

## ELECTROMAGNETIC FORCES

### FINDING THE DIRECTION OF THE FORCE

If a conductor AB (Fig. 11.11) which carries a current is placed at right angles to a magnetic field produced by either a permanent or electromagnet,

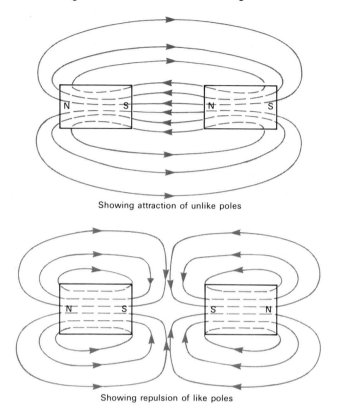

Showing attraction of unlike poles

Showing repulsion of like poles

Fig. 11.10    Magnetic field patterns between two bar magnets

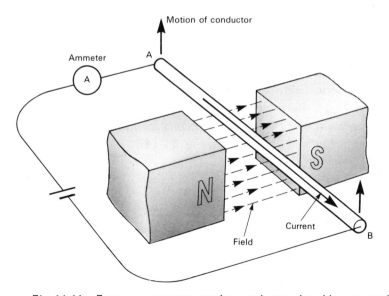

Fig. 11.11    Force on a current-carrying conductor placed in a magnetic field

the conductor will experience a force. This, as in previous cases, is due to the interaction between (a) the field due to the current in the conductor and (b) the other field. If free to move, the conductor will move as shown.

The direction of motion can be found as follows. Fig. 11.12(a) shows a front view of the arrangement with both magnetic fields superimposed on each other, Fig. 11.12(b) shows the resultant field.

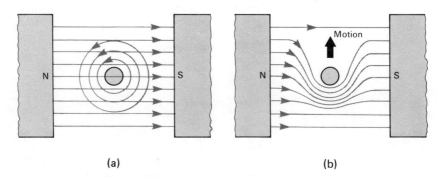

(a)                                (b)

Fig. 11.12    Finding the direction of motion

The fields cancel above the conductor, but add below it. There is, therefore, a strengthening of the field below which we can think of as catapulting the conductor upwards.

**FLEMING'S LEFT-HAND RULE** _____

In Fig. 11.11 we have three quantities whose directions are at right angles: the magnetic field, the current, and the force or motion. These are shown in Fig. 11.13. By using the thumb and first two fingers of the left hand we can represent any two of these quantities and hence find the third. The rule is given below and illustrated in Fig. 11.14:

Fleming's left-hand rule

> First finger points in direction of Field
> seCond finger points in direction of Current
> thuMb gives direction of Motion

The student should be satisfied that both rules stated above are consistent with each other, and use whichever is preferred.

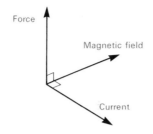

Fig. 11.13    The three quantities, current, field and force are mutually at right angles

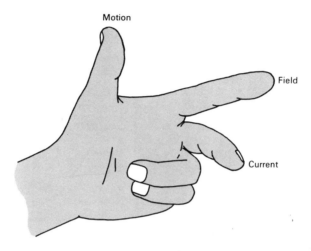

Fig. 11.14    Fleming's left-hand rule for finding direction of motion

## MAGNITUDE OF THE FORCE

When an experiment as in Fig. 11.11 was conducted there are three quantities that can be changed, and which we would expect to influence the force. They are:

(a)    The value of the current, $I$.

(b)    The magnetic flux density, $B$.

(c)    The length of conductor, $L$, at right angles to the magnetic field.

Let us consider these in turn.

(a)    If the current is doubled and everything else kept the same, it is found that the force is doubled. This is indicated in Fig. 11.15.

(b)    If the magnetic flux density is doubled by using a second stronger magnet, and everything else kept the same, it is found that the force is doubled. This is indicated in Fig. 11.16.

(c)    If the length of conductor at right angles to the field could be doubled, and everything else kept the same, it would be found that the force is doubled. This is indicated in Fig. 11.17.

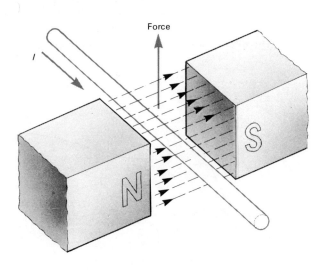

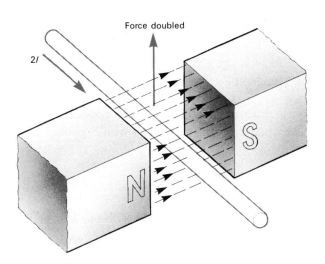

Fig. 11.15    The effect on the force when the current *I* is increased by using a second stronger magnet

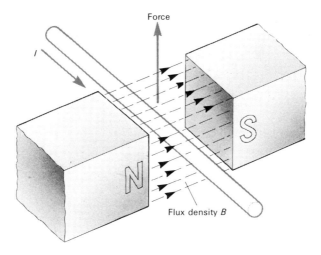

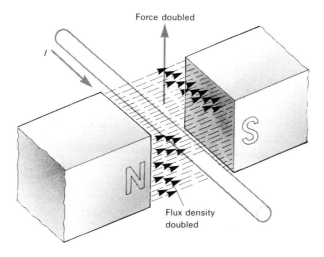

Fig. 11.16    The effect on the force when the flux density *B* is increased

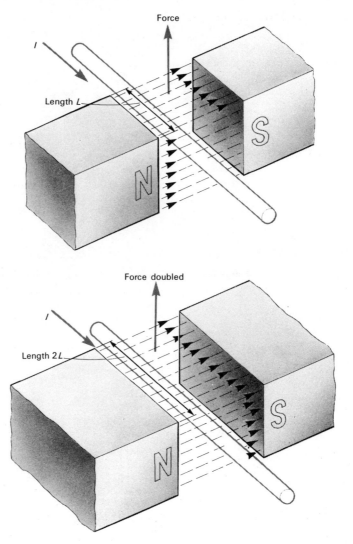

Fig. 11.17    The effect on the force if the length $L$ is increased for the same value of $B$

It can, therefore, be concluded that the force on the conductor is proportional to:

(a)    the current $I$,

(b)    the magnetic flux density $B$,

(c)    the conductor length $L$,

or, in symbol form,

$$\text{Force,} \quad F \propto BLI$$

or, if suitable units are chosen,

$$F = BLI \qquad [11.2]$$

where    $F$ = force in newtons
$B$ = flux density in tesla
$L$ = length in metres
$I$ = current in amperes

Equation [11.2] has applications in both d.c. motors and d.c. meters. It is also used to define the tesla, as follows.

When a wire one metre long carrying a current of one ampere is placed at right angles to a magnetic flux density of one tesla then it experiences a force of one newton.

### WORKED EXAMPLE 3

Find the force on an arrangement such as Fig. 11.11, if the current is 10 A, the length of conductor 0.1 m, the flux $1 \times 10^{-3}$ Wb and the dimensions of the magnetic pole faces are 10 cm $\times$ 2 cm.

### Solution

### Step 1

Find the flux density $B$ using equation [11.1].

The cross-sectional area $A$ is $10 \times 2 \times 10^{-4}$ m$^2$

$$B = \frac{\Phi}{A} = \frac{1 \times 10^{-3}\,\text{T}}{2 \times 10^{-3}}$$

$$B = 0.5\,\text{T}$$

### Step 2

Use $F = BLI$ (equation [11.2])

$$F = 0.5 \times 0.1 \times 10\,\text{N}$$

$$F = 0.5\,\text{N}$$

## THE MOVING-COIL METER

The moving-coil meter is shown in its simplest form in Fig. 11.18. What happens is that when current flows through the moveable coil, the coil experiences a couple tending to rotate it against the return springs. The pointer sets in such a position that the effects of the electromagnetic couple due to the current in the magnetic field balances the mechanical couple of the return springs.

The need for the soft iron core is explained below.

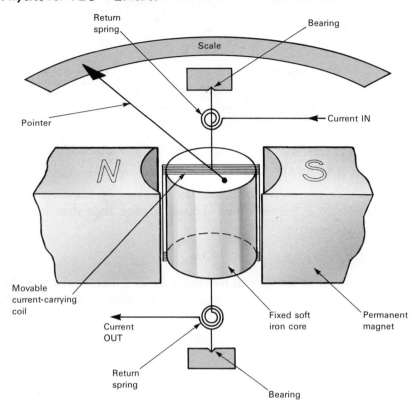

Fig. 11.18    Pointer-type moving-coil meter: simplified practical form

## COUPLE ON A LOOP OF WIRE

The simple arrangement of a current carrying conductor in a magnetic field could be used to measure a current (Fig. 11.11). From equation [11.2] :

$$F = BLI$$

If $B$ and $L$ are known, and $F$ is measured, then $I$ can be found. Unfortunately this geometry is not very convenient. A more useful shape is a current loop (Fig. 11.19). Here the forces on the various sides of the coil are as follows:

AB,    $F = BIb$;    in the direction shown which is that given by Fleming's left-hand rule.

BC,    $F = 0$;    since $I$ and $B$ are parallel.

CD,    $F = BIb$;    in the direction shown.

DA,    $F = 0$;    $I$ and $B$ are parallel.

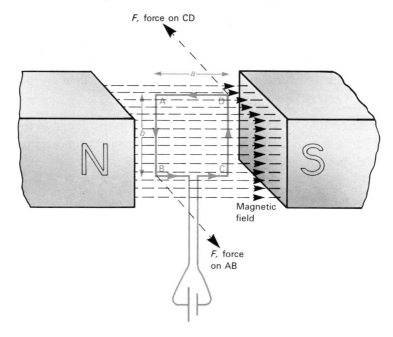

Fig. 11.19    Couple acting on a coil in a magnetic field

The result of two equal and opposite forces is called a couple. The view as seen from the top of the page is Fig. 11.20. A couple causes a turning effect; the turning effect or torque is given by the magnitude of one force multiplied by the distance between them:

$$\text{Torque} = Fa$$
$$= BIba$$
$$= BIA \qquad\qquad [11.3]$$

where $A = ba$ is the area of the coil.

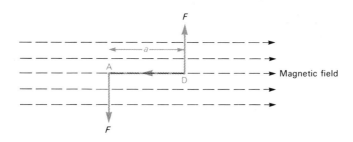

Fig. 11.20    A couple acting on the coil (plan view)

*Notes*

(a)   This can be found by the principle of moments if it is assumed the coil would rotate about a line drawn through the centre of the coil.

(b)   This is only true if the lines of the magnetic field are parallel to sides BC and AD.

This second point is very important because in the design of a d.c. meter the coil turns against a spring giving an elastic torque proportional to the angle turned through, i.e.

$$\text{Elastic torque} = c\theta \qquad\qquad [11.4]$$

$\theta$ here is measured usually in radians (see Chapter 2).

Once the coil turns, however, then equation [11.3] would be incorrect, because $F$ is no longer perpendicular to AB and DC. Some method must be devised to have a field like that in Fig. 11.21. This type of field is impossible since magnetic field lines cannot cross. Something similar can, however, be obtained if one realises it is only at AB and DC that the field must be *radial* (Fig. 11.22). Note the use of the soft iron core.

It can be seen that the field in the annular gap where the coil will turn appears to be radiating into and out of point O, which will be the axis of rotation of the coil. In this case we can equate the electromagnetic and the elastic torques. From equations [11.3] and [11.4]

$$c\theta = BIA$$

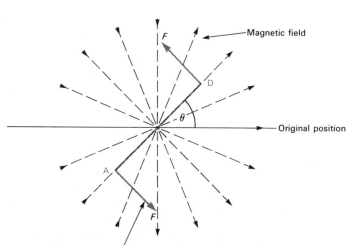

Force $F$ is not straight down the paper as it would be with the field in Fig. 11.20 — making the torque $BIA \cos\theta$

Fig. 11.21   A radial field (theoretical)

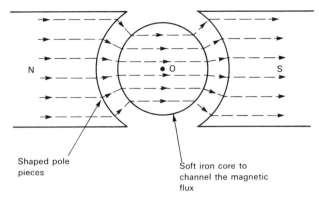

Fig. 11.22    A radial field (practical arrangement)

For a coil with *n* turns this becomes

$$c\theta = BIAn$$

or
$$\frac{\theta}{I} = \frac{BAn}{c}$$
    [11.4]

$\theta/I$ is called the *meter sensitivity*.

An arrangement showing the coil in a meter is given in Fig. 11.23.

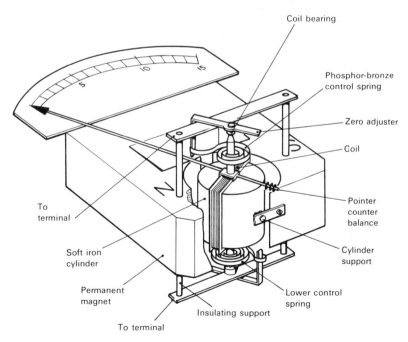

Fig. 11.23    A pointer-type moving-coil milliammeter

*WORKED EXAMPLE 4*

Find the couple on a coil of area $2 \times 10^{-4}$ m² which has 300 turns, and carries a current of 1.5 mA if it is placed parallel to a field of 0.08 T.

*Solution*

$$\text{Electromagnetic couple} = BIAn$$
$$= 0.08 \times 1.5 \times 10^{-3} \times 2 \times 10^{-4} \times 300$$
$$= 7.2 \times 10^{-6} \, \text{N m}$$

Notice how small the value of this couple is; the values of the quantities given are all reasonable for a milliammeter. It can be judged from this that the springs used in meters are quite fragile. (Meters in general don't bounce!)

## FORCE ON CHARGES MOVING IN A MAGNETIC FIELD

Previously the force on a current-carrying conductor in a magnetic field was given by equation [11.2] :

$$F = BLI$$

The current in a wire, however, is due to the flow of electrons which are negatively charged particles. Let us suppose that the origin of the force on the wire is a bulk effect due to the sum of a large number of microscopic forces on individual moving electrons.

Currents flow through metals, and a metal has not only electrons bound to individual atoms, but also free electrons. These in an isolated atom would be the outermost electrons which are the least strongly held. The number of these free electrons varies from metal to metal, but is of the order of one per atom. These 'free' electrons are able to move around within the confines of the lump of metal. When a potential difference is applied across the ends of the metal then the electrons move (Fig. 11.24). Since they are negatively charged their motion is opposite to the conventional current flow. This is because the idea of current was conceived before it was known that a current is actually the flow of electrons.

Conventional current is the movement of positive charges, but since current is actually the flow of electrons (negatively charged) the flows are marked opposite.

Current is rate of flow of charge, i.e.

$$I = \frac{Q}{t} \qquad [11.5]$$

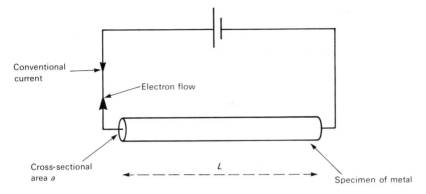

Fig. 11.24    Directions of conventional current and electron flow

The current $I$ is actually the passage of charge $Q$ across the area at some point in a circuit within time $t$. $Q$ is in coulomb. In fact we define the coulomb from equation [11.5]:

> One coulomb of charge passes some point in a circuit in one second if one ampere is flowing.

Later on in the book it will be shown that an applied potential difference actually tries to *accelerate* the electrons, but for electrons in a metal, collisions occur with the metal atoms resulting in the electrons having a drift velocity $v$, proportional to the applied potential difference.

Fig. 11.25 shows the situation in the specimen of metal. To relate $I$ to these moving electrons we will have to use equation [11.5] and find out how many cross the area $a$ in some known time.

If they are moving at speed $v$ then to cross $a$ in one second they must be within $v$ metres since distance travelled equals speed times time. Thus to

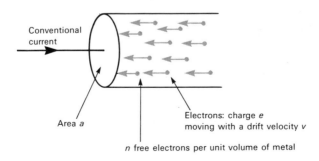

Fig. 11.25    Movement of electrons in a metal

cross $a$ in one second they must be within the volume $a \times v$ (Fig. 11.26). If there are $n$ per unit volume, and each one carried a charge $e^-$, then

$$I = \frac{Q}{t} = \frac{ne^-av}{1}$$

$$I = nave^- \qquad\qquad [11.6]$$

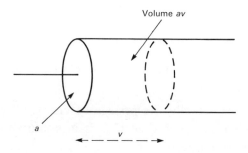

Volume *av*

*a*

*v*

Fig. 11.26    Charge within *av* escapes in one second

If we now consider the effect of putting this specimen in the magnetic field, then from equation [11.2]

$$F = BLI$$

Substituting for $I$ from equation [11.6] gives

$$F = BLnave^-$$

Rearranging gives

$$F = (Be^-v)(naL)$$

Now $aL$ is the total volume of the specimen, so $naL$ is the total number of free electrons in the specimen. Hence $Be^-v$ must be the force on each individual free electron. So:

> Force on an electron moving at speed $v$ at right angles
> to a magnetic field $= Be^-v$      [11.7]

This equation only refers to electrons, but it is true for all moving charges.

Thus for a charge $+q$ moving at speed $v$:

$$\text{Force} = Bqv \qquad\qquad [11.8]$$

The directions are given in Fig. 11.27.

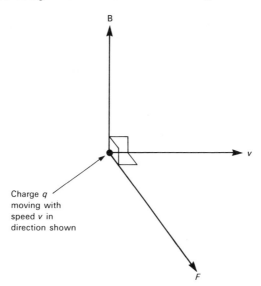

Fig. 11.27    Relative orientation of *v*, *B* and *F*

## WORKED EXAMPLE 5

What will be the force on a charge of $1.6 \times 10^{-19}$ C moving at $10^5$ m s$^{-1}$ in a field of magnetic flux density (*B*) 0.5 T?

*Solution*

From equation [11.8]

$$\text{Force} = Bqv$$
$$= 0.5 \times 1.6 \times 10^{-19} \times 10^5$$
$$= 8.0 \times 10^{-15} \text{ N}$$

Notice how small the force is. A speed of $10^5$ m s$^{-1}$ could only be achieved by an electron in a vacuum. Note that in a metal speeds of about 400 m s$^{-1}$ are more normal. Also note that although the force is very small, the mass of an electron is also very small; so the force would still have a considerable effect (see also Chapter 14).

## EXERCISE 11

1)   Fleming's left-hand rule refers to three quantities. What are they?

2)   A magnetic flux of 50 μWb exists in an air gap formed by magnetic poles of face size  2 cm × 1 cm.  Calculate the flux density in the gap.

3)   The magnetic arrangement of the loudspeaker is shown in Fig. 11.2 (p. 167). Assume that the mean circumference of the air gap is 4 cm and the depth of the gap is 1 cm. If the flux density is 0.1 T, calculate the total flux in the gap. (The width of the gap is assumed to be small.)

4)   If a wire is placed in the gap of Question 2 and you only consider flux being present within the gap, what would be the force on the wire if it carried a current of 2 A and was parallel to the 2 cm sides?

5)   If a single turn coil is placed in the gap of Question 3 and the current is 0.75 A, calculate the force acting on the coil. If when viewed from above the current is clockwise, determine the direction of the force.

6)   A coil of area 0.002 m² is placed at right angles to a uniform magnetic field of flux density 0.15 T. What is the total flux through the coil?

7)   What is the force on an 8 cm wire carrying a current of 250 mA when it is placed in a field of magnetic flux density 0.12 T? You can assume $B$ is perpendicular to the wire.

8)   In Question 7, how could the wire be positioned for the force to be zero?

9)   Calculate the torque on a loop of wire 0.02 m high by 0.015 m wide if it carries a current of 5 mA, and has 40 turns. The magnetic flux density is 0.08 T, and the lines of magnetic flux are parallel to the loop.

10)   State the factors on which the current sensitivity of a moving-coil instrument depend.

11)   If the current sensitivity of a meter is to be increased, which factors would not affect the resistance of the instrument: (a) if the same wire is used for the coil; (b) if different wire is used for the coil?

12)   Sketch an arrangement which would give a radial magnetic field.

13)   A wire of diameter 0.08 mm carries a current of 50 mA. If the current flow is due to electrons (each carrying a charge of $1.6 \times 10^{-19}$ C) moving at $100 \, \text{m s}^{-1}$, how many free electrons are there present in the wire per unit volume?

14)   Calculate the force on a helium nucleus (carrying a charge of $+3.2 \times 10^{-19}$ C) moving in a field of magnetic flux density 0.9 T, and moving at $3 \times 10^5 \, \text{m s}^{-1}$. Assume that $B$ and $v$ are perpendicular.

# 12 ELECTROMAGNETIC INDUCTION AND THE SIMPLE GENERATOR

## GENERATION OF AN EMF FROM A MAGNETIC FIELD _____

Once it had been discovered that an electric current could produce a magnetic field, the obvious next question for physicists was, could a magnetic field produce an electric current in a circuit if that circuit was placed in a magnetic field. The answer is yes, in certain circumstances, and it can easily be demonstrated.

In Fig. 12.1 as the magnet approaches or moves away from the region surrounding the coil a potential difference is indicated on the meter. This potential difference can be used as a primary source of power, and so it is referred to as an electromotive force or e.m.f. The direction (sign) of the e.m.f. is opposite in the two cases. Further experiment indicates that the e.m.f. increases as:

(a)   The speed of motion of the magnet increases.

(b)   The strength of the magnet increases.

(c)   The number of turns on the coil increases.

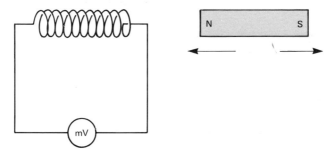

Fig. 12.1   A demonstration of electromagnetic induction. Moving the magnet in and out of the coil gives a deflection on the meter

From this it becomes clearer that what is changing as the magnet moves is the total magnetic flux through the coil. The speed of the magnet alters the rate of change of the flux. Further investigation reveals that the induced e.m.f. in the coil is directly proportional to the rate of change of the flux.

That is:

$$e \propto \frac{\Delta \Phi}{\Delta t}$$

where $e$ is the induced e.m.f., $\Delta \Phi$ is the change in the flux $\Phi$ which occurs in time $\Delta t$.

To understand what is happening as far as the flux is concerned we look at Fig. 12.2. (A single-turn coil is considered.) As the magnet moves away so the flux through the coil from the magnet would fall. However, the e.m.f. that is induced is in such a direction as to produce a current $I$ in the coil which creates a magnetic field (see Chapter 11) in such a direction as to try to keep the total flux constant.

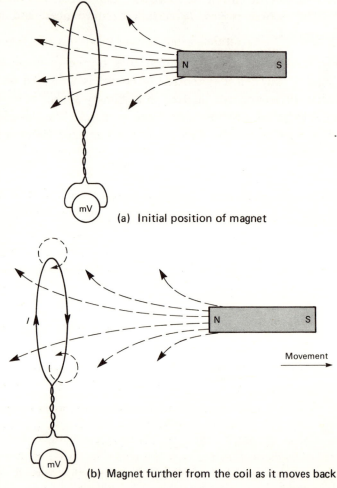

(a) Initial position of magnet

(b) Magnet further from the coil as it moves back

Fig. 12.2   Direction of induced current and its magnetic effect

Thus the induced e.m.f. is in such a direction as to try to keep the total flux constant. That is, the e.m.f. *e* opposes the change of flux. In mathematical terms

$$e = -\frac{\Delta \Phi}{\Delta t}$$     [12.1]

Note the proportionality has become an equality by choosing the units of flux to make it so. Magnetic flux is measured in webers (Wb). The weber is defined by equation [12.1] as that change in total flux through a circuit occurring in one second if an e.m.f. of one volt is generated in the circuit as a result.

The tesla is the unit of flux density, and is the flux density when a total flux of one weber is threading an area of one metre squared.

Equation [12.1] contains a minus sign; what this means is that the induced e.m.f. is in such a direction as to oppose the cause producing it. In Fig. 12.2 as the magnet approaches so an e.m.f. is induced in the coil, therefore, a current flows in the coil. This current causes the end of the coil nearest to the advancing North pole to behave like a north pole, thus repelling the oncoming magnet. This effect of the induced e.m.f. *opposing* the cause producing it is referred to as Lenz's law. Equation [12.1] is called Faraday's law.

Flux through a coil could be made to change with time by other means than moving a magnet. Fig. 12.3 looks at four basic possibilities:

(a)   This is the one we have already discussed. If instead of the magnet simply moving away, the flux through the coil is changed by rotating the coil, then we have the simple generator which follows (see p. 196).

(b)   Here instead of a magnet providing the flux this is done by a second circuit. In industrial a.c. generation part of the output is turned into d.c. (rectified), and this is used to provide the permanent magnetic field in a situation similar to the simple generator.

(c)   This is the situation in a transformer. Transformers are used to step up (make larger) or step down (make smaller) an a.c. voltage.

(d)   Here only one circuit is involved, and one talks of the *self-inductance* of the circuit. In Fig. 12.3(c) one has two circuits, and uses the phrase *mutual inductance* between the circuits.

Since an e.m.f. can be induced in a circuit when the current through it is switched off, this may cause problems in certain circumstances. For example, using a 2 V cell, and a large coil with an iron core, momentary voltages in excess of 50 V can be produced. Thus for switching off large pieces of machinery where a similar situation occurs, capacitors are connected across the switch to stop arcing.

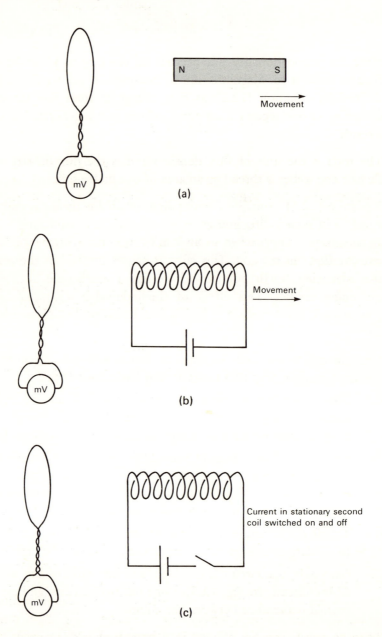

(a)

(b)

(c)

Fig. 12.3 (continued opposite)

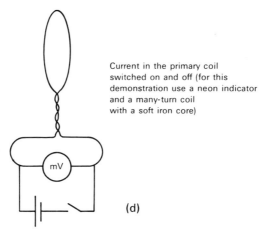

Current in the primary coil switched on and off (for this demonstration use a neon indicator and a many-turn coil with a soft iron core)

(d)

Fig. 12.3    Methods of changing the magnetic flux linking a circuit

## DETERMINATION OF THE STRENGTH OF THE MAGNETIC FIELD BETWEEN THE POLES OF A POWERFUL MAGNET

A method will now be described for finding the magnetic flux density by the use of Faraday's law.

Consider a coil of $n$ turns and an area $A$. If it is placed at right angles to the lines of a magnetic field of flux density $B$, then the total flux through the coil is

$$\Phi = BAn$$

If this total flux is removed in time $\Delta t$, then using Faraday's law

$$e = -\frac{\Delta(BAn)}{\Delta t}$$

Since only $B$ is to change, this becomes

$$e = -An\frac{\Delta B}{\Delta t} \qquad [12.2]$$

The coil used is called a search coil (see Fig. 12.4(a)). It is placed in the region where the flux density $B_{initial}$ is to be found, and is then removed quickly to a region where the flux density $B_{final}$ is effectively zero. If the unknown flux density is that between the poles of a magnet (Fig. 12.4(a)) then the search coil is placed inside the gap, and then removed away from the influence of the magnet. An alternative to complete removal of the search coil is to rotate it through $90°$ as shown in Fig. 12.4(b) when the coil presents no area for the flux to pass through. A $180°$ rotation would double the flux change.

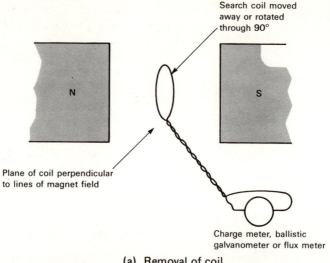

Search coil moved away or rotated through 90°

Plane of coil perpendicular to lines of magnet field

Charge meter, ballistic galvanometer or flux meter

**(a)  Removal of coil**

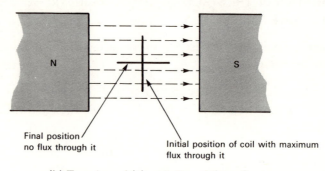

Final position no flux through it

Initial position of coil with maximum flux through it

**(b) Top view of (a): rotation of the coil**

Fig. 12.4    Use of a search coil to determine magnetic flux density

There is a choice of measuring instrument. With a *flux meter* the time period of removal is unimportant, and the flux change through the coil is measured directly. With a *charge meter* or *ballistic galvanometer* time can be eliminated mathematically as shown below from equation [12.2]:

$$e = -An\frac{\Delta B}{\Delta t}$$

Rearranging gives          $e\Delta t = -An\Delta B$

Substituting          $e = IR$    (Ohm's law)

in the above gives

$$IR\Delta t = -An\Delta B$$

$I$ is the current and $\Delta t$ is the time, and $I\Delta t = Q$.

The definition of the coulomb is the charge passing some point in a circuit in one second if one ampere is flowing (amps $\times$ seconds = charge passed in coulombs). $R$ is the resistance of the circuit. Thus

$$R\Delta Q = -An\Delta B$$

This equation indicates that if the search coil changes position or orientation to give a flux change through it, then the magnetic flux density change $\Delta B$ can be found if $\Delta Q$ the charge passed is measured. It is assumed that the resistance of the search coil/ballistic galvanometer circuit is known, and that the coil area and number of turns are known.

$$\Delta B = -\frac{R}{An}\Delta Q \qquad [12.3]$$

$\Delta Q$ is measured directly on a chargemeter, or can be found from the initial deflection on a ballistic galvanometer:

$$\Delta B = B_{\text{initial}} - B_{\text{final}}$$

$B_{\text{initial}}$ is the magnetic flux density required, and $B_{\text{final}}$ is zero. Thus the magnetic flux density in the gap can be found.

## WORKED EXAMPLE 1

A ballistic galvanometer indicates that a charge of $5\,\mu C$ has passed round a circuit of resistance $500\,\Omega$. The circuit contains a coil of $100$ turns of area $10^{-4}\,m^2$. This charge passes when the coil is suddenly removed from between the poles of a magnet. What is the magnetic flux density between the poles? It is assumed that the coil is moved to a region of low magnetic field compared to the initial position.

## Solution

Using equation [12.3] and since $\Delta B = B$, the flux density between the poles, then

$$B = \frac{R\Delta Q}{An}$$

$$= \frac{500 \times 5 \times 10^{-6}}{100 \times 10^{-4}}$$

$$= 0.25\,T$$

*Note.* The minus sign in equation [12.3] indicates only the direction of charge flow, so this sign is omitted above.

## THE SIMPLE ALTERNATING CURRENT (AC) GENERATOR ___

The generator consists of $n$ turns each of area $A$ rotating in a permanent magnetic field. This is the simplest case to consider (as mentioned previously), and the arrangement is shown in Fig. 12.5. As the coil rotates the area through which flux from the magnet passes changes. At the position indicated in Fig. 12.6, the flux through the coil is

$$\Phi = BAn \sin \theta$$

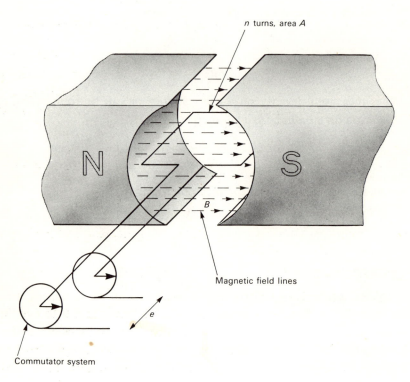

*n* turns, area *A*

Magnetic field lines

Commutator system

Fig. 12.5   Principle of the a.c. generator

$A \sin \theta$ is the area presented by the area $A$ to the magnetic flux from the permanent magnet. By Faraday's law the e.m.f. $e$ produced is given by equation [12.1]:

$$e = -\frac{\Delta\Phi}{\Delta t}.$$

In differentiated terms this is written $e = -\dfrac{\mathrm{d}\Phi}{\mathrm{d}t}.$

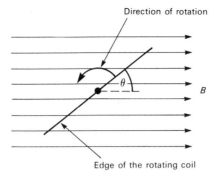

Fig. 12.6    Instantaneous position of coil (side view of Fig. 12.5)

The flux at an instant $t$ seconds after passing through the horizontal is $\Phi = BAn \sin \theta$. From equation [2.2] we can write $\theta = \omega t$ where $\omega$ is the angular speed of rotation of the coil. If the frequency of rotation of the coil is $f$, then $\omega = 2\pi f$ since in one rotation the coil sweeps out an angle $2\pi$ radians (see Chapter 2). Thus $\theta = \omega t = 2\pi ft$. Hence

$$e = -\frac{d\Phi}{dt} = -\frac{d}{dt}(BAn \sin \theta) = -\frac{d}{dt}(BAn \sin 2\pi ft)$$

$$e = -BnA\, 2\pi f \cos (2\pi ft)^* \qquad\qquad [12.4]$$

The quantity $BnA\, 2\pi f$ is a constant for the generator, and can be written as $e_0$. Hence the output of the generator is

$$e = -e_0 \cos (2\pi ft) \qquad\qquad [12.5]$$

The plot of equation [12.5] is shown in Fig. 12.7, and below it the actual coil positions at the instant of time indicated. It can be seen that the output voltage or e.m.f. is a maximum when the *rate of flux change* through the coil is a maximum, not when the flux linkage is a maximum. So the maximum e.m.f. occurs when the coil is parallel to the magnetic field. The output voltage is described as *alternating*, and produces an *alternating current* (a.c.) when connected across a resistance or other circuit components. The word 'alternating' implies changes from positive to negative, and back repeatedly; i.e. the current flows first in one direction and then in the opposite direction.

### WORKED EXAMPLE 2

A coil of 200 turns each of area $6 \times 10^{-4}\,\text{m}^2$ rotates at 2400 r.p.m. in a field of 0.08 T. What will be the maximum output voltage from the system?

*This differentiation is covered in the TEC Mathematics II unit: see *Mathematics for Technicians: New Level II* by A Greer and G W Taylor (also published by Stanley Thornes, 1982).

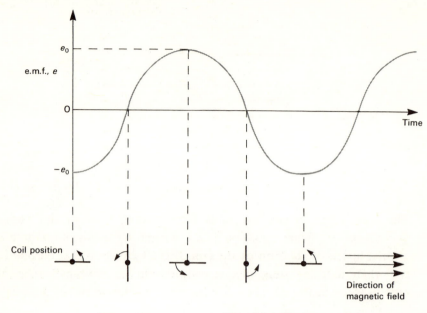

Fig. 12.7    Output e.m.f. and coil position

*Solution*

From equation [12.4]  $e = -BnA2\pi f \cos(2\pi ft)$.

This is a maximum when $\cos(2\pi ft)$ is a maximum. The maximum value of the cosine function is 1, and this occurs when the rotating coil is parallel to the magnetic field. Hence the maximum value of $e$ is $e_0$.

$$e_0 = -BnA2\pi f$$

Ignoring the minus sign since we are only interested in magnitudes:

$$e_0 = 0.08 \times 200 \times 6 \times 10^{-4} \times 2\pi f$$

$f$ the frequency is put into this expression in Hz or rotations per second. We convert from r.p.m. to r.p.s. or Hz by dividing by 60 (seconds in a minute). Hence

$$f = \frac{2400}{60} = 40 \, \text{Hz}$$

$$e_0 = 0.08 \times 200 \times 6 \times 10^{-4} \times 2\pi \times 40$$

$$= 2.41 \, \text{V}$$

## MUTUAL INDUCTION

We have seen that current in a coil will produce magnetic flux. If we have an iron-cored toroid, as in Fig. 12.8, then there will be magnetic flux as shown.

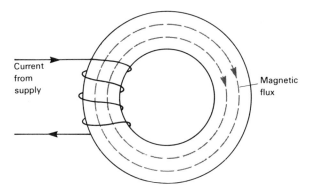

Fig. 12.8    Flux in a toroid

If now a second coil is wound on the toroid, and the coils are connected as shown in Fig. 12.9, then the following can be observed:

(a)    At the moment when switch S is closed the meter will deflect momentarily showing that a voltage is generated across the second coil.

(b)    While S is closed, despite the fact that current flows in coil 1 and, therefore, magnetic flux exists in the ring, there is no voltage across coil 2.

(c)    When the switch S is opened, there is again a momentarily deflection of the meter. This time in the opposite direction from that when the switch was closed.

The phenomenon observed is known as *mutual induction*. Although there is no electrical contact between the two coils, it is possible through the medium of the magnetic flux to generate or induce a voltage in the second coil. It will be noted that:

(a)    Voltage is only generated if the flux changes. When the flux is constant (constant current in coil 1) there is no voltage across coil 2.

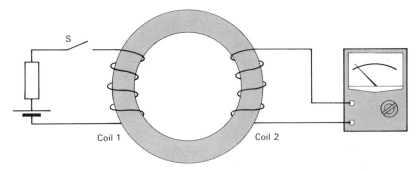

Fig. 12.9    An experiment to show mutual induction

(b)    When the flux in coil 2 increases from zero to a certain value, as occurs when the switch is closed, the generated voltage is of a certain polarity or direction. When the flux decreases, as occurs on switching off, the generated voltage is of the opposite polarity.

The crucial factor with regard to the generated voltage is the *flux linkage* with coil 2. The flux linkage is the product of the number of turns $N_2$ in coil 2 and the flux $\Phi_2$ through it. Thus

$$\text{Flux linkages} = N_2\Phi_2$$

## MAGNITUDE OF GENERATED VOLTAGE

The generated voltage is proportional to the *rate of change* of flux-linkages, and if correct units are chosen we can write

Generated voltage in coil 2, $e_2$ = Rate of change of $(N_2\Phi_2)$

$$= N_2 \times \text{Rate of change of } \Phi_2$$
(since $N_2$ is constant)

In calculus notation, this becomes

$$e_2 = \frac{\mathrm{d}}{\mathrm{d}t}(N_2\Phi_2)$$

or, since $N_2$ is constant,

$$e_2 = N_2\frac{\mathrm{d}\Phi_2}{\mathrm{d}t}$$

If $N_2$ is made very large — that is, we have a large number of turns on coil 2 — the generated voltage can be very great. In fact, much greater than the battery voltage connected across coil 1. This principle is made use of in the

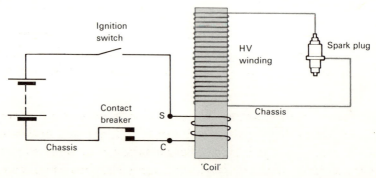

Fig. 12.10    The principle of an ignition system in a motor car

motor-car ignition system which produces a high voltage across the sparking plug (of the order of several thousand volts) from a 12 V battery. Because the voltage is very high the current is very small. The system is shown in Fig. 12.10.

### WORKED EXAMPLE 3

The magnetic flux linking a coil of 500 turns increases linearly from $10 \times 10^{-4}$ Wb to $50 \times 10^{-4}$ Wb over a time interval of 100 ms. Calculate the generated voltage across the terminals of the coil during this time interval.

*Solution*

We use the equation $e = N \dfrac{d\Phi}{dt}$.

*Step* 1: Find $\dfrac{d\Phi}{dt}$.

$$\frac{d\Phi}{dt} = \frac{\text{Change of flux}}{\text{Time taken for that change}}$$

$$= \frac{(50-10) \times 10^{-4} \text{Wb}}{100 \times 10^{-3} \text{s}}$$

$$= \frac{40 \times 10^{-4} \text{Wb}}{10^{-1} \text{s}} = 40 \times 10^{-3} \text{Wb s}^{-1}$$

*Step* 2:

$$e = N \times \frac{d\Phi}{dt}$$

$$= 500 \times 40 \times 10^{-3} \text{V}$$

$$= 20 \text{V}$$

## THE TRANSFORMER

The transformer is a device for use in a.c. systems. Its operation depends on mutual induction as described in the previous sections. In principle it consists of two separate coils wound on a common iron core, such as shown in Fig. 12.11.

One coil, known as the *primary* coil is connected to an a.c. supply, the other coil, known as the *secondary* coil can feed power to a load or consumer. If the secondary voltage is larger than the primary voltage it is known as a *step-up* transformer. If the secondary voltage is less than the primary, then it is called a *step-down* transformer.

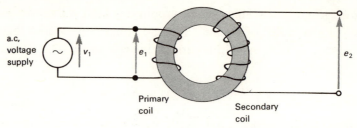

Fig. 12.11    The principle of the transformer

## TRANSFORMER ACTION

The voltage applied to the primary coil is sine-wave a.c., and hence the current in the primary coil will be sinusoidal. This will give rise to a sinusoidal flux which will link with the secondary coil. Thus the flux is *continually* changing according to a sine-wave pattern, and this will generate a voltage across the secondary.

We have $\qquad e_2 = N_2 \times$ Rate of change of $\Phi$

or $\qquad e_2 = N_2 \times$ Rate of change of sine-wave flux

Now since the rate of change (or differential) of a sine-wave is a cosine wave, then the voltage $e_2$ will be of a cosine form which is, of course, identical in shape to a sine wave. This is one of the reasons for using sine-wave supplies, so that the waveform is not changed or distorted in 'passing through' a transformer. The standard symbol for a transformer is shown in Fig. 12.12.

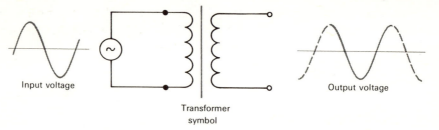

Fig. 12.12    The symbol of a transformer and its input and output waveforms

## VOLTAGE RATIO

For the secondary coil, we have

$$e_2 = N_2 \times \text{Rate of change of } \Phi$$

For the primary coil, since the same flux $\Phi$ links with $N_1$ as well, we have

$$e_1 = N_1 \times \text{Rate of change of } \Phi$$

but $e_1 = v_1$, since there is no voltage drop in the leads connecting them.

Hence: $$v_1 = N_1 \times \text{Rate of change of } \Phi$$

Dividing, we have

$$\frac{e_2}{v_1} = \frac{N_2}{N_1}$$

or $$\frac{\text{Secondary voltage}}{\text{Primary voltage}} = \frac{N_2}{N_1}$$ [12.6]

so, the ratio of the secondary to primary voltage is the same as the 'turns-ratio' of the transformer.

### WORKED EXAMPLE 4

For the safe use of power tools a step-down transformer is required to change the mains voltage of 240 V to 24 V. What turns-ratio is required?

### Solution

From equation [12.6] we have

$$\frac{N_2}{N_1} = \frac{\text{Secondary voltage}}{\text{Primary voltage}} = \frac{24}{240} = \frac{1}{10}$$

Therefore, the secondary winding should have one-tenth of the number of turns on the primary.

### WORKED EXAMPLE 5

Two transformers are linked together as shown in Fig. 12.13. It is required that the output voltage of the second transformer should be 400 V. Calculate the number of secondary turns required for the second transformer.

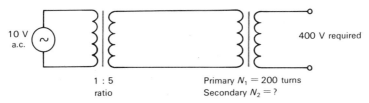

| 10 V a.c. | | | 400 V required |

1 : 5
ratio

Primary $N_1$ = 200 turns
Secondary $N_2$ = ?

Fig. 12.13  Information for Worked Example 5

### Solution

The voltage output of the first transformer will be $10 \times 5 = 50$ V. For the second transformer we then have, from equation [12.6]

$$\frac{N_2}{N_1} = \frac{400}{50} = \frac{8}{1}$$

and $$N_2 = 8 \times N_1 = 8 \times 200 = 1600 \text{ turns}$$

## VOLTAGE AND CURRENT RATIOS

Well designed transformers are highly efficient, their efficiency can easily exceed 95%. In other words, very little power is lost *in* the transformer itself. Let us assume there are no losses. Then

$$\text{Power output} = \text{Power input}$$

For the transformer shown in Fig. 12.14, if the load is resistive, we have

$$\text{Power input} = V_1 I_1 \text{ watts}$$

$$\text{Power output} = \text{Power in load} = V_2 I_2 \text{ watts}$$

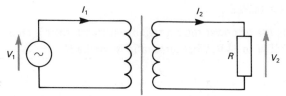

Fig. 12.14    Voltage and current ratios

Therefore, if

$$\text{Power output} = \text{Power input}$$

$$V_2 I_2 = V_1 I_1$$

or

$$\frac{V_2}{V_1} = \frac{I_1}{I_2} \qquad [12.7]$$

This means that if we step-up the *voltage* we step-down the *current*, and vice versa.

### WORKED EXAMPLE 6

A transformer supplies 400 W to a resistance load connected to its secondary. Assuming there are no losses in the transformer, and that its turns ratio is 1:2, calculate the primary and secondary currents if the primary voltage is 50 V.

### Solution

Since there are no losses:

$$\text{Input power} = \text{Output power} = 400 \text{ W}$$

Therefore

$$V_1 I_1 = 400 \text{ W} \quad \text{(from equation [12.7])}$$

or

$$I_1 = \frac{400 \text{ W}}{50 \text{ V}} = 8 \text{ A}$$

Since the step-up turns ratio is 1:2, the voltage output will be increased two-fold and the current decreased by a half.

Therefore:                    Output current = 4 A

This can be checked:

$$Output\ voltage\ =\ 2 \times 50\ V\ =\ 100\ V$$

So, since

$$V_2 I_2\ =\ 400\ W$$

$$I_2\ =\ \frac{400\ W}{100\ V}\ =\ 4\ A$$

**EXERCISE 12** _____

1)   Calculate the e.m.f. generated in a circuit containing a single-turn coil through which there is a flux change of 0.5 mWb in 0.02 s.

2)   Recalculate Question 1 for a coil of 200 turns.

3)   Calculate the e.m.f. generated when a coil of area $2 \times 10^{-3} m^2$ and 150 turns is rotated from a position where it is at right angles to the flux to a position where it is parallel to the flux in 20 ms. The coil is between the pole pieces of a magnet where the magnetic flux density is 0.02 T.

4)   A charge meter indicates a charge of 2 $\mu$C has passed round a circuit of total resistance 2 k$\Omega$ when a coil of 50 turns each of area $0.2 \times 10^{-4} m^2$ is suddenly removed from a position between the poles of a magnet. What is the magnetic flux density between the poles if the initial position is at right angles to the flux, and the final one in a region of negligible field?

5)   A coil of 150 turns, each of area $2 \times 10^{-3} m^2$ is rotated at 800 r.p.m. about an axis at right angles to a magnetic field of magnetic flux density 0.12 T. Calculate the maximum voltage generated.

6)   A circular coil of radius 0.080 m and 50 turns makes 2500 r.p.m. in a field of magnetic flux density 0.018 T. Calculate the maximum generated voltage.

7)   Calculate the voltage generated across a coil of 2000 turns, when the flux through the coil falls from 0.003 Wb to 0.001 Wb in 50 ms.

8)   What ratio of turns is required in a transformer to step 250 volts down to 12 V?

9)   A power station has an output of 5 MW at 440 V. The power is transformed up to 120000 V for transmission into the national grid. Calculate: (a) the input current,  (b) the turns ratio.

10)   For a step-up transformer, which of the following statements is/are incorrect:

   (i)   The secondary voltage is less than the primary voltage.
  (ii)   The secondary current is less than the primary current.
 (iii)   The secondary power exceeds the primary power.
  (iv)   The secondary turns exceed the primary turns.

# 13 ELECTRIC FIELD

## INTRODUCTION

It is common experience that when a sweater made of a man-made fibre is removed, and the lower garment is also of a man-made fibre, a spark or an electrical discharge of some kind occurs. What has happened is that a separation of *charge* has occurred. One garment has ended up with a surplus of electrons, and the other with a deficit. We say that they are charged positive if there is a deficit, and negative if there is a surplus of electrons.

Any charge affects the region of space surrounding it. We say that a charge distribution causes an *electric field* in the region surrounding the charge. This electric field shows up when a second charge enters the region because it experiences a force. The force depends on the magnitudes of the charges and their signs.

Like charges repel, unlike charges attract. *Thus we can describe an electric field as a region where a charged body experiences a force.*

## ELECTRIC FIELD STRENGTH

The size of the force will depend on the magnitudes of the charge and field in which it is placed. If a charge $+q$ coulombs (the unit of charge) experiences a force of $F$ newtons at some point, then we define the electric field strength $E$ at the point by the relationship

$$E = F/q \qquad \text{[13.1]}$$

The unit of electric field strength would appear to be $N\,C^{-1}$ or newtons per coulomb, but this is not the unit in common use as will be seen later. The unit normally written is $V\,m^{-1}$.

The coulomb is a unit derived from the ampere: if you consider a wire through which an electric current of one ampere is flowing, then a charge of one coulomb passes any point in that wire in one second.

## WORKED EXAMPLE 1

Calculate the electric field strength in the region where a 2 mC charge experiences a force of 4 N.

*Solution*

From equation [13.1] $E = F/q$.

We have $F = 4\,\text{N}$, $q = 2 \times 10^{-3}\,\text{C}$

Hence

$$E = \frac{4}{2 \times 10^{-3}} = 2000$$

The electric field strength is 2000 V m$^{-1}$.

Although electric field is a very useful concept, the more usual industrial situation is that a potential difference is established between two points and a charge separation results. To establish the connexion between electric fields and potential differences, we use the force that the charge experiences.

Consider the simple case of a charge $+Q$ on a small sphere in space. In this case, if we move a small charge $+q$ about in the region surrounding it, and measure the *direction* of the resulting force, we can plot out the electric field lines. The result is as shown in Fig. 13.1. This situation, although the simplest possible, is not a very useful one or a normal industrial charge distribution. A more normal one is shown in Fig. 13.2, where there is a charge separation, and the charges are spread over metallic conductors. Although the situation appears more complex, the electric field is seen to be much simpler. This electric field is *uniform*. This means that it is constant over the whole of the region between the conductors. The electric field is constant in value, and always in the same direction.

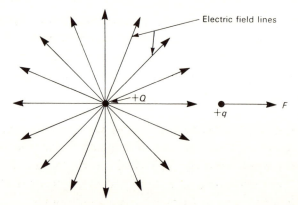

Fig. 13.1    Electric field due to a charge $+Q$ on a small metal sphere. The arrows represent the direction in which the force acts on charge $+q$

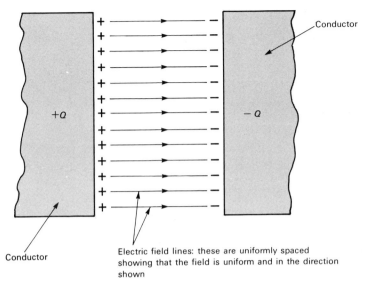

Electric field lines: these are uniformly spaced
showing that the field is uniform and in the direction
shown

Conductor

Fig. 13.2 A uniform electric field

If a charge of $+q$ coulombs ($q$ is a small fraction of a coulomb) was placed on a small metal sphere anywhere between the two surface charges, then there would be a force on it pushing it toward the negatively charged surface. The movement would be spontaneous if the charge $+q$ was free to move. However, to move the charge $+q$ in the opposite direction would require work to be done. The work required is given by the product of the force times the distance moved for a constant force.

Thus the nearer the charge $+q$ is to the positively charged plate, the more work it could do; i.e. the greater its potential energy, the greater its capacity to do work, the higher its electrical potential. We say the positive plate is at a higher potential and talk of the *potential difference* between the plates. The potential difference is measured by the work required to move $+1$ *coulomb of charge* from the negative to the positive surface.

If the work required is 1 joule, then we define the potential difference as 1 volt.

The definition applies to any two regions or points between which a potential difference exists. Strictly, since a movement of charge would itself alter the potential at a point, it should be a small charge. In general then the potential difference between two points is defined as the work done per unit charge in moving a (small) charge from one point to the other.

From the above definition we see that if the potential difference is $\Delta V$, then the work $\Delta W$ required to move a small charge $+q$ from the point of lower potential to that of the higher potential is given by

$$\Delta W = q\Delta V \qquad [13.2]$$

## ELECTRIC FIELD STRENGTH AND POTENTIAL GRADIENT

Suppose we require to move a charge $+q$ a small distance $\Delta x$ toward the positive surface as shown in Fig. 13.3. Since the electric field is uniform then a constant force $F$ is required. Work must be done and from Chapter 4

Work done = Force × Distance moved

$$W = F\Delta x$$

From equation [13.1] $E = F/q$ so if $q$ is $+1$ coulomb we can write $E = F$, so in this case $\Delta W = E\Delta x$.

From equation [13.2] and the definition of the volt

$$\Delta W = q\Delta V = 1\Delta V = \Delta V$$

Combining $\Delta W = \Delta V$ and $\Delta W = E\Delta x$ we find

$$\Delta V = E\Delta x \quad \text{or} \quad E = \frac{\Delta V}{\Delta x}$$

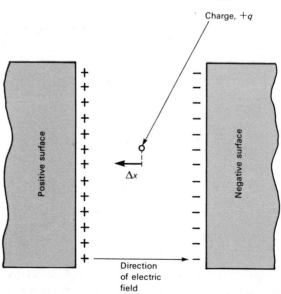

$\Delta x$ is a small distance towards the positive surface

Fig. 13.3   Movement of charge $+q$ by $\Delta x$ towards positive surface

The normal practice is to call displacements positive when they are away from a positive charge. Hence

$$E = -\frac{\Delta V}{\Delta x}$$    [13.3]

Thus $E$ is the negative of the potential gradient $\frac{\Delta V}{\Delta x}$. Note that the direction of $E$ is from positive to negatively charged surface (Fig. 13.3). Note that since $V$ is in volts and $x$ is in metres, then the electric field strength $E$ is in volts per metre.

### WORKED EXAMPLE 2

Fig. 13.4(a) shows two parallel metallic plates with a potential difference between them. They are separated by a distance $d$ of 0.020 m. Calculate the electric field strength in the region between the plates, and the force acting on a charge of $+5$ nC in this region.

*Solution*

From equation [13.3] $E = -\frac{\Delta V}{\Delta x}$.

Since the potential difference is 1000 V and $\Delta x = 0.020$ m

$$E = -\frac{\Delta V}{\Delta x} = -\frac{1000}{0.020} = -50\,000 \text{ V m}^{-1}$$

The electric field strength is 50000 V m$^{-1}$ and the negative sign indicates that this field is in the opposite direction to increasing potential. This can be seen in Fig. 13.4(b), which shows the variation of potential between the plates.

The force $F$ acting on the 5 nC charge is given by equation [13.1] $F = qE$

where $q = 5 \times 10^{-9}$ C and $E = 50000$ V m$^{-1}$

Hence
$$F = 5 \times 10^{-9} \times 50000$$
$$F = 25 \times 10^{-5} \text{ N}$$

## ACCELERATION OF CHARGED PARTICLES

If the charge $+q$ is placed near the positively charged surface of Fig. 13.3 and then released, it would experience a force which continues to be felt all the way across the gap between the two surfaces. By the time it reached the negatively charged surface it would have acquired energy. This energy would be equal to the work required to push it back to the positive plate. From equation [13.2]

$$\text{Energy acquired} = \Delta W = q\Delta V$$

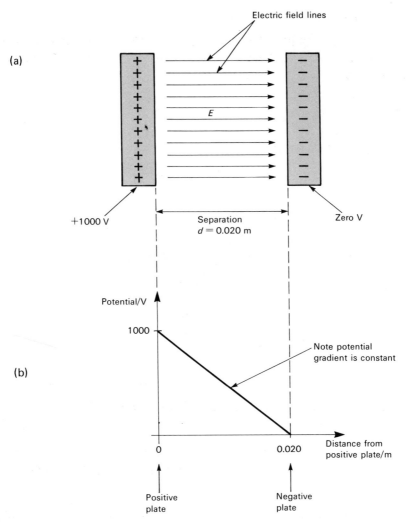

Fig. 13.4   (a) Information for Worked Example 2.
           (b) Variation of potential between the plates in (a)

The energy it has when it reaches the negative plate is all kinetic energy ($\frac{1}{2}mv^2$) where $m$ is the mass on which the charge $+q$ resides, and $v$ is the velocity acquired on reaching the negative surface. Hence

$$\tfrac{1}{2}mv^2 = q\Delta V \qquad\qquad [13.4]$$

This method of accelerating charges has been used in research for accelerating electrons and protons, and is the method used to produce an accelerated beam of electrons in a television tube. The combination of electron source and accelerator is often referred to as an electron gun (see Chapter 14).

*WORKED EXAMPLE 3*

An electron (charge $1.6 \times 10^{-19}$ C and mass $9.1 \times 10^{-31}$ kg) is released at some point in a vacuum, and accelerated through a potential difference of 1000 V. Calculate the kinetic energy it acquires, and its final speed.

*Solution*

From equation [13.2] $\Delta W = q \Delta V$.

Since $q = 1.6 \times 10^{-19}$ C and $\Delta V = 1000$ V.

$$\Delta W = 1.6 \times 10^{-19} \times 1000$$
$$= 1.6 \times 10^{-16} \text{ J}$$

This is the kinetic energy acquired, and hence

$$\Delta W = \tfrac{1}{2} m v^2$$
$$1.6 \times 10^{-19} = \tfrac{1}{2} \times 9.1 \times 10^{-31} \times v^2$$

Rearranging gives

$$v^2 = \frac{3.2}{9.1} \times 10^{15}$$

or

$$v = 1.9 \times 10^7$$

The speed acquired is $1.9 \times 10^7$ m s$^{-1}$.

---

**EXERCISE 13** _____

1)  Calculate the force on an electron (charge $1.6 \times 10^{-19}$ C) in an electric field of $2.3 \times 10^4$ V m$^{-1}$.

2)  Calculate the acceleration that an electron has when placed in an electric field of $4.2 \times 10^5$ V m$^{-1}$ (electron charge $1.6 \times 10^{-19}$ C and electron mass $9.1 \times 10^{-31}$ kg).

3)  What is the electric field between the plates of the parallel plate capacitor in Fig. 13.2 if the plate separation is 0.5 mm and the p.d. between the plates is 250 V?

4)  If in Question 3 a proton is accelerated across the gap, what would be the gain in energy of the proton (charge on the proton is $1.6 \times 10^{-19}$ C)?

5)  An alpha particle is accelerated through a potential difference of 20000 V, what would be its final speed if it can be considered to have started from rest? (The mass of the alpha particle is $6.64 \times 10^{-27}$ kg, and its charge is $3.2 \times 10^{-19}$ C.)

6)  A corona breakdown occurs in a gas at $3.1 \times 10^6$ V m$^{-1}$. Use this fact to determine the maximum p.d. that can occur between two metal plates 4.0 mm apart in the gas.

# ELECTRONS AND IONS

## THE ELECTRON

An electron is a particle of mass $m_e$ equal to $9.1 \times 10^{-31}$ kg with a charge $e$ equal to $-1.6 \times 10^{-19}$ C.

## DEFLECTION BY AN ELECTRIC FIELD

A uniform electric field exists between two parallel plates connected to a battery as in Fig. 14.1. If the potential difference between the plates is $V$ and their separation $d$ then the electric field strength $E = V/d$ (see Chapter 13). Fig. 14.1 shows an electron in this electric field. The electron will experience an upwards force because it is attracted to the positive plate. This is of magnitude $F$ given by (see Chapter 13)

$$F = eE$$

$$= -1.6 \times 10^{-19} \times E$$

The negative sign indicates that the force is in the opposite direction to the field.

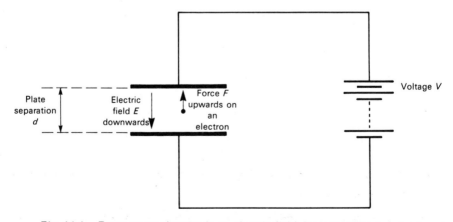

Fig. 14.1   Force on an electron in an electric field. Note that the drawing is not to scale: the plates should be shown much closer together for a uniform field

214

In Fig. 14.2 an electron initially moving horizontally enters the electric field region. As a result of the force acting on it the electron follows the path shown. In the electric field region the force acts parallel to the field which causes the path of the electron to be parabolic.

Throughout this chapter we shall neglect the weight of the electron. This is because the force due to gravitational attraction is negligible compared to the forces due to electric, and magnetic fields.

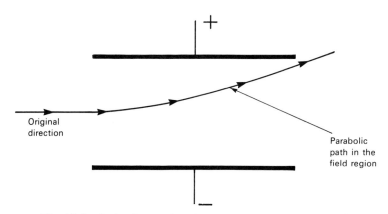

Fig. 14.2   Path of a moving electron in an electric field. Note that the path of the electron is straight when outside the field region

## WORKED EXAMPLE 1

Refer to Fig. 14.1. If the potential difference between the plates is 300 V and their separation 6 cm calculate: (a) the electric field strength between the plates, (b) the force acting on an electron in this region. Calculate also the weight of an electron.

*Solution*

(a)
$$E = V/d = \frac{300}{0.06}$$
$$= 5000 \text{ V m}^{-1}$$

(b)
$$F = eE = 1.6 \times 10^{-19} \times 5000$$
$$= 8.0 \times 10^{-16} \text{ N}$$

Taking the acceleration due to gravity $g$ as $9.8 \text{ m s}^{-2}$, then the weight of an electron is given by

$$\text{Weight} = mg = 9.1 \times 10^{-31} \times 9.8$$
$$= 8.9 \times 10^{-30} \text{ N}$$

Note that the weight is much less than (b).

# DEFLECTION BY A MAGNETIC FIELD

An electron has a charge so that a moving electron constitutes an electric current. A moving electron thus experiences a force in a magnetic field and hence is deflected. The magnitude of the force $F$ is given by (see Chapter 11)

$$F = Bev$$

where    $v$ = velocity of the electron
$e$ = charge on electron
$B$ = intensity of magnetic field

This assumes the electron moves at right angles to the field. The direction of the force is given by Fleming's left-hand rule. It is perpendicular to both the electron motion and field direction. In Fig. 14.3 an electron initially moving horizontally enters the magnetic field region. The force it experiences is always perpendicular to its motion so that it moves in a circular path, as shown in Fig. 14.3.

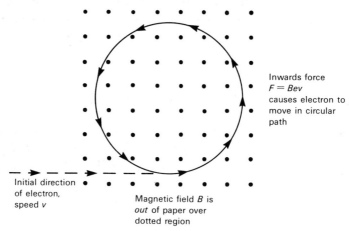

Inwards force
$F = Bev$
causes electron to
move in circular
path

Initial direction
of electron,
speed $v$

Magnetic field $B$ is
*out* of paper over
dotted region

Fig. 14.3    Path of a moving electron in a magnetic field

## WORKED EXAMPLE 2

(a)    Find the magnitude of the force acting on an electron moving with speed $8 \text{ Mm s}^{-1}$ at right angles to a magnetic field of intensity 0.4 mT.

(b)    Calculate the radius of the circular path described by the electron.

*Solution*

(a)                         $F = Bev$

$= 0.4 \times 10^{-3} \times 1.6 \times 10^{-19} \times 8 \times 10^{6}$

$= 5.12 \times 10^{-16} \text{ N}$

(b)   The radius of the path is found by equating the force *Bev* to the inwards force necessary to keep the electron moving in a circle (see Equation 2.6, p. 21).

That is
$$F = Bev = \frac{m_e v^2}{r}$$
[14.1]

where     $m_e$ = mass of electron
$v$ = velocity of electron
$r$ = radius of circular path

Rearranging gives

$$r = \frac{m_e v}{eB} = \frac{9.1 \times 10^{-31} \times 8 \times 10^6}{1.6 \times 10^{-19} \times 0.4 \times 10^{-3}}$$
$$= 0.114$$

The radius of the circular path is 0.114 m.

# DETERMINATION OF ELECTRON MASS AND CHARGE _____

It is difficult to directly measure the mass $m_e$ of an electron. Usually the mass is found by doing two experiments — one to measure the specific charge $e/m_e$ and one to measure the electron charge $e$. Combining the two results gives $m_e$.

# DETERMINATION OF SPECIFIC CHARGE $e/m_e$ _____

Several methods exist for measuring $e/m_e$ and two different types are now described. In these methods a magnetic field, of known intensity $B$ is used to deflect a beam of electrons of fixed velocity $v$. Equation [14.1] gives

$$Bev = \frac{m_e v^2}{r}$$

Rearranging gives

$$\frac{e}{m_e} = \frac{v}{rB}$$
[14.2]

so that we must measure $v$, $r$ and $B$ in each experiment.

### (i) ELECTRIC AND MAGNETIC DEFLECTION – THOMSON'S METHOD _____

Fig. 14.4(a) shows a modern version of Thomson's apparatus. The electron gun produces a thin horizontal beam of electrons, all having the same velocity. The beam trajectory can be seen on the fluorescent screen. The electrons may be deflected in a vertical direction either electrostatically, by applying

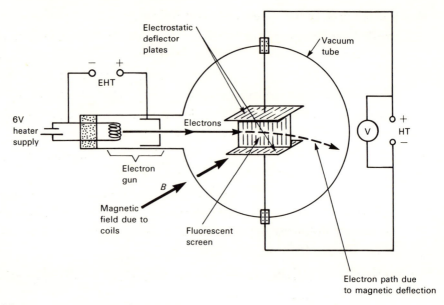

(a) Thomson's apparatus (modern form)

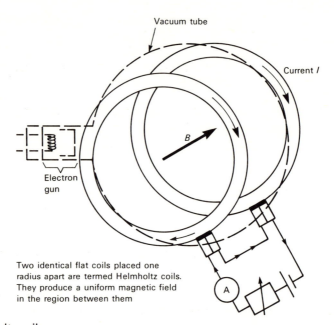

Two identical flat coils placed one radius apart are termed Helmholtz coils. They produce a uniform magnetic field in the region between them

(b) Helmholtz coils

Fig. 14.4    Thomson's determination of $e/m_e$

a potential difference across the deflector plates, or magnetically by passing a current through a pair of coils called *Helmholtz coils* (see Fig. 14.4(b)). The electric and magnetic deflections are arranged to be in opposite directions.

The procedure for determining $e/m_e$ is as follows:

(a)    With both electric and magnetic fields set to zero the initial path of the electron beam is noted.

(b)    A magnetic field is now applied by passing a suitable current $I$ through the Helmholtz coils so that a suitable deflection of the beam is observed.

(c)    The *radius r* of the electron path is calculated from the new position of the beam (using manufacturer's literature).

(d)    The *magnetic field intensity B* is calculated from the current $I$ (using manufacturer's literature).

(e)    The *velocity v* of the electrons is measured by applying a potential difference $V$ to the electrostatic deflector plates until the beam returns to its initial position in (a). The electric field strength $E = V/d$, where $d$ is the separation of the plates.

When the two deflections cancel out the forces due to the magnetic and electric effects must be equal. That is when

$$eE = Bev \qquad [14.3]$$

or

$$v = \frac{E}{B} \qquad [14.4]$$

Thus the velocity of the electrons is found. Use of equation [14.2] gives $e/m_e$.

In practice a more accurate value for the specific charge may be found using magnetic deflection only, as below.

## (ii) MAGNETIC DEFLECTION – FINE BEAM TUBE

This apparatus is shown in Fig. 14.5. The electron gun emits a thin beam of electrons in the vertical direction. The beam is made visible by having a small amount of hydrogen gas inside the tube.

In this experiment the electron beam is deflected into a complete circular path in the tube, using a known magnetic field. The procedure is as follows:

(a)    A known voltage $V_g$ is applied to the electron gun. The velocity $v$ acquired by the electrons is, from Chapter 13, given by

$$\tfrac{1}{2}m_e v^2 = eV_g$$

Rearranging gives

$$v^2 = 2\left(\frac{e}{m_e}\right)V_g \qquad [14.5]$$

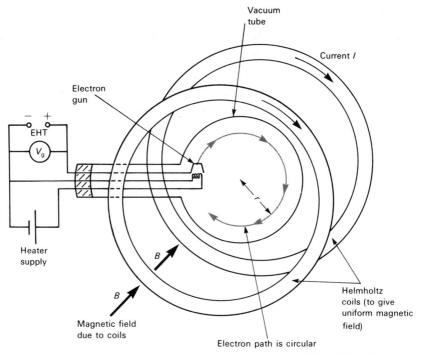

Fig. 14.5    Fine beam tube

(b)    A magnetic field is applied, perpendicular to the electron beam, by passing a current through the pair of coils (Helmholtz coils — see Fig. 14.4(b)). The current is increased until the electron beam forms a complete circle in the tube (see Fig. 14.5). The current value is used to calculate the magnetic field intensity (from the manufacturer's literature).

(c)    The radius of the circle formed by the electron beam is measured. Note that we cannot calculate the electron velocity $v$ from equation [14.5] since we do not know $e/m_e$. We square equation [14.2] to give

$$\left(\frac{e}{m_e}\right)^2 = \frac{v^2}{r^2 B^2}$$

Substituting for $v^2$ from equation [14.5] into this expression gives

$$\left(\frac{e}{m_e}\right)^2 = \frac{v^2}{r^2 B^2} = 2\left(\frac{e}{m_e}\right)\frac{V_g}{r^2 B^2}$$

or                $$\frac{e}{m_e} = \frac{2V_g}{r^2 B^2}$$                [14.6]

Typical results are given in the following worked example.

## WORKED EXAMPLE 3

For a particular fine beam tube the magnetic field intensity $B$, in teslas, due to a current $I$ in the Helmholtz coils is given by

$$B = 4.2 \times 10^{-3} \times I$$

It was found that when a potential difference of 300 V is applied to the electron gun that a current of 0.29 A passing through the Helmholtz coils results in the electron beam forming into a circle of radius 4.8 cm. Use the data to find the specific charge on the electron.

### Solution

Equation [14.6] gives

$$\frac{e}{m_e} = \frac{2V_g}{r^2 B^2}$$

where    $V_g = 300$ V

$r = 4.8 \times 10^{-2}$ m

$B = 4.2 \times 10^{-3} \times 0.29$

$\quad = 1.22 \times 10^{-3}$ T

Thus    $$\frac{e}{m_e} = \frac{2 \times 300}{(4.8 \times 10^{-2})^2 \times (1.22 \times 10^{-3})^2}$$

$$= 1.75 \times 10^{11} \text{ C kg}^{-1}$$

The accepted value is $1.76 \times 10^{11}$ C kg$^{-1}$.

## (iii) MAGNETIC DEFLECTION – THOMSON TUBE

The apparatus used is shown in Fig. 14.4 but magnetic deflection only is used. The procedure is as follows:

(a)  A known voltage $V_g$ is applied to the electron gun so that electrons are emitted with velocity $v$ given by equation [14.5].

(b)  A magnetic field is applied by passing a current through the Helmholtz coils, until the electron beam is suitably deflected.

(c)  The magnetic field intensity is calculated from the current in the coils.

(d)  The radius $r$ of the electron path is calculated from the new position of the beam.

Typical results are given below.

## WORKED EXAMPLE 4

The following results were obtained in an experiment using a version of the Thomson's apparatus to measure $e/m_e$ by magnetic deflection.

Voltage on electron 'gun' $V_g$ = 3000 V

Magnetic field intensity $B$ = 0.87 mT

Radius of electron path $r$ = 0.21 m

Calculate the specific charge on an electron.

*Solution*

Equation [14.6] gives

$$\frac{e}{m_e} = \frac{2V_g}{r^2B^2} = \frac{2 \times 3000}{(0.21)^2 \times (0.87 \times 10^{-3})^2}$$

$$= 1.8 \times 10^{11} \, \text{C kg}^{-1}$$

## MILLIKAN'S EXPERIMENT

Bodies are electrically charged if they have an excess or a deficit of electrons compared to the neutral state. Millikan's experiment, first reported in 1909, showed that electric charge comes in 'packets' of $1.6 \times 10^{-19}$ C. This basic unit is the charge on an electron and is denoted by $e$.

A simplified version of Millikan's experiment is now described. Suppose a small sphere of known mass $m$ and carrying electric charge $q$ is introduced into a uniform electric field region (see Fig. 14.6). The direction and magnitude

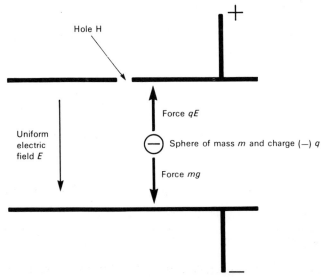

Fig. 14.6    Principle of Millikan's apparatus

$E$ of the field is now adjusted until the sphere remains stationary. This occurs when

Force downwards due to weight = Force upwards due to electric field

or
$$mg = qE$$

or
$$q = \frac{mg}{E}$$

If all quantities on the right-hand side are known we can measure $q$. A simulation of Millikan's experiment can be performed using Latex or polystyrene spheres and their charge measured.

In Millikan's original experiment oil droplets were used. A fine sprayer was used to form tiny spherical oil droplets above a hole H in the upper plate (see Fig. 14.6) and occasionally a droplet fell through H and into the field region. The droplet was charged in the process of its formation or by ionising the air around it using a beam of X-rays. The mass of an individual droplet is unknown and has to be determined by a separate experiment. This involves switching off the electric field and measuring the speed of the droplet as it falls. The equation of fluid dynamics enable the mass of the droplet to be found.

Repeated experiments, with different droplets show that, although the charge on different droplets may differ, the charge on a droplet is always a whole number times the value $1.6 \times 10^{-19}$ C. We conclude that this is the basic unit of charge and is the charge carried by one electron.

### WORKED EXAMPLE 5

In a Millikan-type experiment with six different oil droplets the following values for charge on the droplets were found:

| Droplet | A | B | C | D | E | F |
|---|---|---|---|---|---|---|
| Charge/$10^{-19}$ C | 8.0 | 9.6 | 4.8 | 17.6 | 6.4 | 3.2 |

What is the basic unit of charge according to this experiment?

### Solution

We note that the charges are all multiples of $1.6 \times 10^{-19}$ C as follows:

| Droplet | A | B | C | D | E | F |
|---|---|---|---|---|---|---|
| Charge/$(1.6 \times 10^{-19}$ C) | 5 | 6 | 3 | 11 | 4 | 2 |

We conclude that the basic unit of charge is $1.6 \times 10^{-19}$ C. Note that in practice droplets with charge $1e$ are only rarely obtained.

# THE ATOM

The simplest model of an atom is composed of a nucleus made up of protons and neutrons surrounded by a number of electrons equal to the number of protons (see Fig. 14.7). The approximate masses and charges are given in Table 14.1).

TABLE 14.1   Constituents of the atom

| Particle | Mass/a.m.u.* | Charge/C |
|----------|--------------|----------|
| Proton   | 1            | $+1.6 \times 10^{-19}$ |
| Neutron  | 1            | 0 |
| Electron | $\dfrac{1}{1840}$ | $-1.6 \times 10^{-19}$ |

*1 a.m.u. = 1 atomic mass unit
$\quad = \frac{1}{12}$ of mass of a $^{12}_{6}$C atom
$\quad = 1.66 \times 10^{-27}$ kg

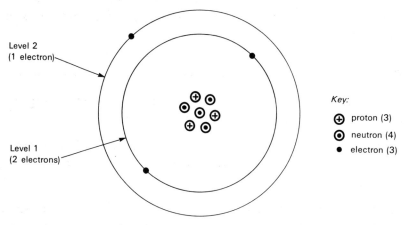

Level 2
(1 electron)

Level 1
(2 electrons)

Key:
⊕  proton (3)
◉  neutron (4)
●  electron (3)

Fig. 14.7   Simplified model of an atom (of lithium-7)

# ISOTOPES

All atoms of the same *element* have nuclei with the same number of *protons*. The atomic number $Z$ is the number of protons in the nucleus so that $Z$ identifies the element. (For example, all carbon atoms have 6 protons and all uranium atoms have 92 protons.)

The number of neutrons in a nucleus is called the neutron number $N$. This can vary for a given element. If two nuclei have the same number of protons but a different number of neutrons they are called *isotopes* of the given element. For example all carbon nuclei have 6 protons but carbon can exist with 6, 7 or 8 neutrons.

The mass number $A$ equals the sum of the proton and neutron numbers.

That is

$$A = N + Z$$

We denote a given nucleus by the symbol $^A_Z X$, where X is the chemical symbol of the element. The isotopes of carbon have the following symbols:

$^{12}_6 C$ which has 6 protons and 6 neutrons

$^{13}_6 C$ which has 6 protons and 7 neutrons

$^{14}_6 C$ which has 6 protons and 8 neutrons

Isotopes cannot be separated by chemical means because the electron structure and properties, which determine chemical behaviour, are determined by the number of protons. Thus isotopes have identical chemical behaviour, and in order to separate them we must use physical techniques such as the mass spectrometer, described below.

## IONS

An ion is an atom or group of atoms which have become charged, positively or negatively, through losing or gaining electrons.

In sodium chloride (common salt) a sodium atom loses 1 electron to become a sodium ion ($Na^+$). This electron is transferred to the chlorine atom which becomes a chlorine ion ($Cl^-$).

In the mass spectrometer, ions are produced from neutral gas atoms as shown in Fig. 14.8. The gas enters a low-pressure region in which an electrical discharge is established between an anode and a cathode with a potential difference of about 10 000 V. The discharge causes electrons to be stripped from the atoms leaving positive ions. These are attracted to the cathode and some pass through to give a beam of positive ions.

## THE MASS SPECTROMETER

This instrument is capable of identifying the presence of isotopes. It is used in industry for a variety of purposes — for example, the measurement of

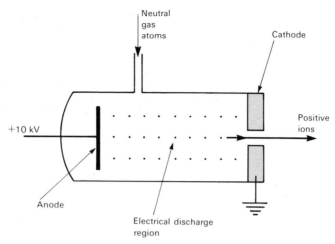

Fig. 14.8    A source of gaseous ions

molecular masses of oils, the measurement of isotope ratios in the nuclear industry, and the measurement of high vacuum and leak detection in low-pressure work.

A diagram of a Bainbridge-type mass spectrometer is shown in Fig. 14.9. A source I emits ions with a range of velocities. The velocity selector permits only those with a given known velocity to pass through the slit S at the entrance to the high-vacuum chamber. These ions pass into a uniform magnetic field region, of known magnitude, which means that the ions follow a circular path as shown. The radius of this circle depends upon the mass of the ions and thus their mass is measured. For ions of a given element it is possible to detect the presence of isotopes, since different isotopes have circular paths of different radius.

**THE VELOCITY SELECTOR** _____

This ensures that all ions entering the slit S have the same velocity. In the velocity selector region of Fig. 14.9 the ions may be deflected in the directions shown either electrostatically, by applying a voltage to the metal deflector plates, or magnetically by a magnetic field directed at right angles as shown. The electric and magnetic deflections are arranged to be in opposite directions. Suppose an ion with charge $+q$ and velocity $v$ passes straight through. The forces due to electric and magnetic effects must be equal. That is

$$qE = B_1 qv$$

[14.7]

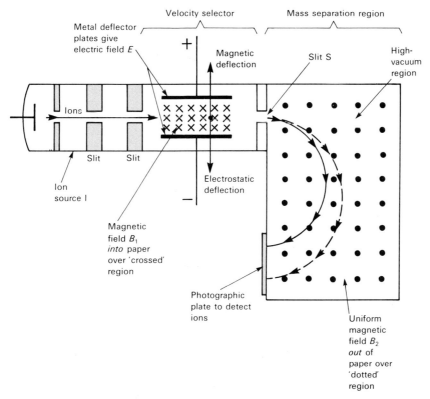

Fig. 14.9   The Bainbridge mass spectrometer

where $E$ is the electric field strength and $B_1$ the magnetic field intensity in the velocity selector region. Rearranging equation [14.7] we see that only ions with velocity $v$ given by

$$v = \frac{E}{B_1} \qquad [14.8]$$

can pass through slit S.

Note that *all ions* with velocity $v$, *irrespective of their mass*, will follow straight paths through the velocity selector.

### WORKED EXAMPLE 6

A beam of singly charged ions of neon are emitted by the source of a mass spectrometer and enter the velocity selector region. If the electric field strength is $50\,\text{kV m}^{-1}$ and the magnetic field intensity is $0.20\,\text{T}$ in the velocity selector region, calculate the speed of ions which are undeviated.

*Solution*

From equation [14.8]

$$v = \frac{E}{B_1} = \frac{50 \times 10^3}{0.2}$$

$$= 2.5 \times 10^5$$

Ions with speed $2.5 \times 10^5 \, \text{m s}^{-1}$ are undeviated.

## SEPARATION OF MASSES

Positive ions with a single known velocity enter the mass separation region shown in Fig. 14.9. This region has a uniform magnetic field $B_2$ acting outwards so that the ions are deflected into circles as shown. Suppose the ions all have the same charge $+q$. Thus the radius $r$ of the circular path followed by an ion of mass $m$ is given, from equation [14.1], by

$$B_2 q v = \frac{m v^2}{r}$$

or
$$r = m \left( \frac{v}{q B_2} \right). \qquad [14.9]$$

Since the bracketed term in equation [14.9] has a fixed value then $r$ depends on $m$, so that ions of different mass are separated into circular paths of different radii. Thus the mass $m$ of a given ion type may be found.

The ions form dark lines, which are images of slit S, on the photographic plate as shown in Fig. 14.10. In practice it is possible to read masses directly from the photographic plate by calibrating the mass spectrometer using ions of known mass which give standard lines.

Fig. 14.10   Appearance of photographic plate

### WORKED EXAMPLE 7

Singly charged neon ions with a speed of $2.5 \times 10^5 \, \text{m s}^{-1}$ enter the mass separation region of a Bainbridge mass spectrometer. A magnetic field of intensity 0.30 T deflects the ions into circular paths. Calculate the radius of the path for ions of (a) Ne-20 and (b) Ne-22. (Assume 1 a.m.u. $= 1.66 \times 10^{-27} \, \text{kg}$, electron charge $e = 1.60 \times 10^{-19} \, \text{C}$.)

*Solution*

The ions are singly and positively charged. Thus $q = +1.6 \times 10^{-19}$ C.

(a)  A neon 20 ion has mass $m = 20$ a.m.u. $= 3.32 \times 10^{-26}$ kg. From equation [14.9] with $m = 3.32 \times 10^{-26}$, $v = 2.5 \times 10^5$, $q = 1.6 \times 10^{-19}$ and $B_2 = 0.30$, then the radius $r$ of the path is given by

$$r = \frac{mv}{qB_2} = \frac{3.32 \times 10^{-26} \times 2.5 \times 10^5}{1.60 \times 10^{-19} \times 0.30}$$

$$= 0.173 \text{ m}$$

(b)  From equation [14.9] we see that for ions having the same velocity and charge then $r$ is directly proportional to $m$. For Ne-20, $m = 20$ and $r = 0.173$. For Ne-22, $m = 22$ and suppose $r = R$.

Thus
$$r \propto m$$

so
$$\frac{R}{0.173} = \frac{22}{20}$$

so
$$R = 0.190 \text{ m}$$

The radius of the path for Ne-22 is 0.190 m.

# IONISATION ENERGY

Fig. 14.7 shows a simplified energy level diagram of an atom, in this case lithium-7. It has an equal number of protons and electrons so that it is a neutral atom. The electrons orbit the nucleus and only certain orbits are allowed, so the electrons have certain well-defined energies. Work must be done to remove an electron from an atom because attractive forces bind it to the atom. An electron which is a long way away from an atom is designated zero energy. Thus electrons in an atom have certain well-defined negative energy values, as shown in Fig. 14.11.

Some of the allowed energy levels will not contain electrons. When the electrons are in their lowest possible levels, the atom is said to be in its ground state. When higher energy levels are occupied we say the atom is in an excited state.

The energy required to just remove an electron completely from an atom is called the *ionisation energy*. The first ionisation energy is the energy needed to just remove an electron from the highest energy level when the atom is in its ground state. It is the energy needed to just remove the uppermost electron shown in Fig. 14.11. We can also define a second ionisation energy which is the energy needed to remove a second electron and so on.

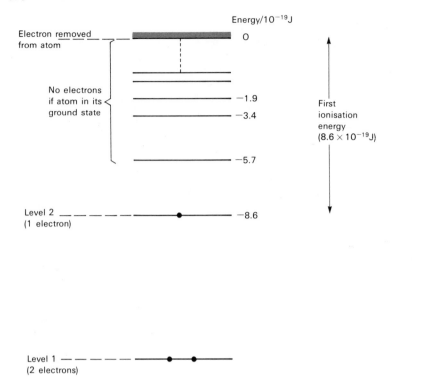

Fig. 14.11    Electron configuration for a lithium atom in its ground state

## MEASUREMENT OF IONISATION ENERGY

A circuit diagram of a commercially available apparatus to measure the ionisation energy of the atoms of a gas is shown in Fig. 14.12. Electrons emitted by the heated cathode K are accelerated by the potential difference $V_g$ between K and the grid G. As shown in Chapter 13, the electrons acquire kinetic energy $eV_g$ just as they reach grid G. If such an electron collides with a gas atom, then, provided that the incident electron has a certain minimum energy, a bound electron may be knocked out of the atom.

When ionisation of the gas atoms occurs there will be a sudden increase in the number of free electrons and positive ions in the valve.

This is detected by observing the current $I$ flowing through the milliammeter. So the experiment consists of raising the voltage on the variable supply and observing the value of accelerating voltage $V_g$ at which current $I$ suddenly increases.

Alternatively a graph of accelerating voltage $V_g$ versus current $I$ may be

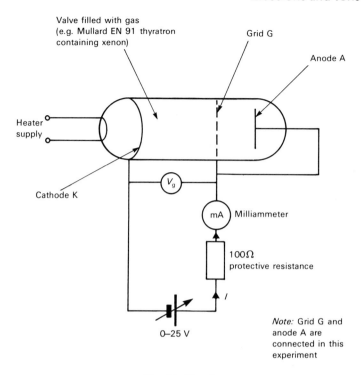

Fig. 14.12    Apparatus to measure ionisation energy

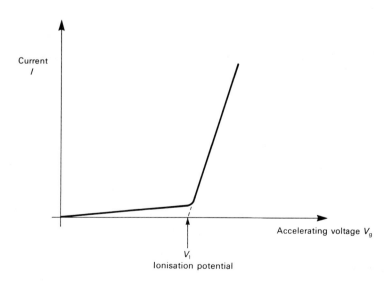

Fig. 14.13    Current versus voltage for ionisation energy

obtained. Typical results are shown in Fig. 14.13. Ionisation first occurs at voltage $V_I$, called the *ionisation potential*, and the ionisation energy of the atoms is thus $eV_I$.

### WORKED EXAMPLE 8

An experiment was performed using the apparatus of Fig. 14.12, with a xenon-filled tube. It was found that as the accelerating voltage $V_g$ was gradually increased there was a significant increase in the current $I$ when $V_g$ reached 12.1 V. Calculate the ionisation energy of xenon.

*Solution*

$$\text{Ionisation energy} = eV_I$$

where
$$e = \text{Electron charge} = 1.6 \times 10^{-19}\,\text{C}$$
$$V_I = \text{Ionisation potential} = 12.1\,\text{V}$$

Thus
$$\text{Ionisation energy} = 1.6 \times 10^{-19} \times 12.1$$
$$= 19.4 \times 10^{-19}\,\text{J}$$

Note that the collision between electron and atom is *inelastic*: that is *kinetic energy is not conserved*. Some of the kinetic energy of the bombarding electron is converted into doing work to remove the bound electron from the atom.

### EXERCISE 14

(Assume where necessary:
Electron charge $e = 1.6 \times 10^{-19}\,\text{C}$
Mass of electron $m_e = 9.1 \times 10^{-31}\,\text{kg}$)

1) (a) An electron is projected at right angles into a region containing a uniform electric field of magnitude $8.0\,\text{kV m}^{-1}$. Calculate the force on the electron.

   (b) When a uniform magnetic field of magnitude 0.50 mT is now applied perpendicular to the electron motion it is found that the electric and magnetic forces cancel each other. Calculate the speed of the electron. Draw a diagram to show the relative orientation of electron motion and electric and magnetic fields.

2) Refer to Fig. 14.14 which shows an electron initially at rest in the electric field region between two metal plates. Which of diagrams A, B or C represents the upwards force on the electron as it moves from plate X to plate Y?

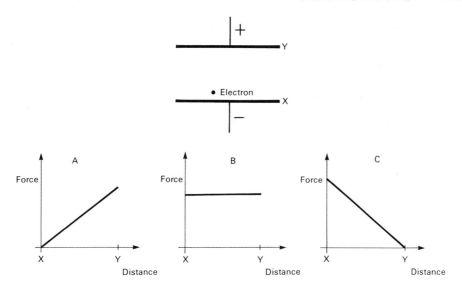

Fig. 14.14    Information for Question 2

3) An electron with speed 12 Mm s⁻¹ moves at right angles to a uniform magnetic field of intensity 1.5 mT. Explain why the electron moves in a circular path and calculate the radius of this path.

4) The following results were obtained using a fine beam tube to determine the specific charge on an electron. The voltage applied to the electron gun was 200 V and application of a transverse magnetic field of intensity 0.78 mT caused the electrons to be deflected into a circular path of radius 0.060 m. Use these data to calculate the specific charge on an electron.

5) Explain the principle of one experiment to determine the specific charge of an electron. No apparatus details are required, but full details of the essential mathematics and observations should be given.

6) Electrons are accelerated from rest through a potential difference of 2000 V and then enter a uniform magnetic field region of intensity 1.0 mT, the direction of which is at right angles to the motion of the electrons. Find the radius of the resultant electron path.

7) Suppose that a Millikan-type experiment gave the following values for charge on the oil droplets:

| Charge/$10^{-19}$ C | 6.4 | 3.2 | 12.8 | 3.2 | 9.6 |
|---|---|---|---|---|---|

What conclusions could you come to regarding the basic unit of charge. Why do you think this differs from the accepted value?

8)   A narrow beam of ions of differing speeds leaves the source of a mass spectrometer and passes into the velocity selector region. If a fixed transverse magnetic field of intensity 0.30 T is applied calculate:

(a)   the strength of the electric field needed to enable ions travelling at 0.50 Mm s$^{-1}$ to be selected;

(b)   the speed of ions which pass undeviated when an electric field of strength 60 kV m$^{-1}$ is applied.

9)   Give a brief account of the function and principle of operation of the following parts of the mass spectrometer:

(a)   ion source;

(b)   velocity selector region;

(c)   mass separation region.

10)   In a particular mass spectrometer the velocity selector region is inoperative so that a beam of singly charged ions with a range of speeds enters the mass selector region. Describe and explain what you would expect to *see* on the photographic plate used to detect the ions.

11)   A beam of singly charged chlorine atoms with a fixed speed of 0.40 Mm s$^{-1}$ passes into the mass separation region of a mass spectrometer. If a magnetic field of intensity 0.50 T is applied calculate the radius of the paths described by (a) Cl-35 and (b) Cl-37 ions. (Assume 1 a.m.u. = 1.66 × 10$^{-27}$ kg.)

12)   The first ionisation energy of lithium is 8.6 × 10$^{-19}$ J. Calculate the ionisation potential of lithium.

13)   For a gaseous mercury atom in its ground state, all electron levels above − 10.6 eV are empty. Calculate the first ionisation energy of mercury.

14)   Calculate the minimum speed of an electron which can cause ionisation of a xenon gas atom. (Ionisation potential of xenon = 12.1 V.)

# ANSWERS

## EXERCISE 1

1) 25 m
2) 139 s, 1.93 km
3) (a) 3.33 s   (b) 15 m
4) (a) 11.33 s   (b) 187 m
5) (a) $0.3\,\mathrm{m\,s^{-2}}$   (b) $0.38\,\mathrm{m\,s^{-2}}$
   (c) 45 m
6) Velocity a vector, speed a scalar
7) 130 s, 167 m
8) $7.5\,\mathrm{m\,s^{-1}}$, 2660 s
9) 14.7 km
10) $208\,\mathrm{m\,s^{-1}}$
11) $0.1\,\mathrm{m\,s^{-2}}$
12) $6.7\,\mathrm{m\,s^{-2}}$
13) (a) $4\,\mathrm{m\,s^{-2}}$   (b) 32 m
14) $2.8\,\mathrm{m\,s^{-1}}$
15) 35 m
16) 1.00 MN
17) (a) 196 N   (b) 123 m
18) 50 kg
19) 489 N
20) 5 N
21) 3.15 kN
22) 6.9 kN
23) (a) 400 N   (b) 3000 N
24) (a) 4000 N   (b) 1000 N, 600 N
25) (a) 789 N   (b) 331 N
26) (a) retarding at $0.65\,\mathrm{m\,s^{-2}}$
    (b) moving at uniform speed
    (c) accelerating at $0.65\,\mathrm{m\,s^{-2}}$

## EXERCISE 2

1) $6000\,\mathrm{rev\,min^{-1}}$
2) $5.65\,\mathrm{m\,s^{-1}}$
3) $6.28\,\mathrm{m\,s^{-1}}$
4) (a) $9\,\mathrm{m\,s^{-2}}$   (b) 45 N
5) 2.6 kN
6) 359 N, 241 N
7) $17\,\mathrm{m\,s^{-1}}$

8) (a) $4.4 \times 10^{16}\,\mathrm{rad\,s^{-1}}$
   (b) $2.2 \times 10^{6}\,\mathrm{m\,s^{-1}}$
   (c) $9.7 \times 10^{22}\,\mathrm{m\,s^{-2}}$
   (d) $8.7 \times 10^{-8}\,\mathrm{N}$
9) (a) $5.27 \times 10^{-13}\,\mathrm{N}$   (b) $2.0 \times 10^{-9}\,\mathrm{N}$

## EXERCISE 3

1) $7.5\,\mathrm{m\,s^{-1}}$
2) $3\,\mathrm{m\,s^{-1}}$ to the left
3) 5 N
4) 15 N
5) 90 N
6) $2.33\,\mathrm{m\,s^{-1}}$
7) $0.33\,\mathrm{m\,s^{-1}}$ in direction of 2000 kg truck
8) $14\,\mathrm{m\,s^{-1}}$
9) $2.98\,\mathrm{m\,s^{-1}}$
10) $1.58\,\mathrm{m\,s^{-1}}$
11) $807\,\mathrm{km\,s^{-1}}$ in forward direction
12) $200\,\mathrm{m\,s^{-1}}$ in original direction

## EXERCISE 4

1) 5 kJ
2) 2400 J
3) (a) 480 kJ   (b) 655 kJ
4) 23.5 kJ
5) (a) 90 J   (b) 160 J   (c) 227 kJ
6) 1.96 kJ
7) (a) 800 J   (b) 16.3 m
8) 617 kJ, 6.29 m
9) 1.47 kJ, $22.1\,\mathrm{m\,s^{-1}}$, 14.0 m
10) 26.1 m
11) 9.0 mJ, 0.46 m
12) (a) 25.1 J   (b) $19.8\,\mathrm{m\,s^{-1}}$
    (c) $-11.9\,\mathrm{m\,s^{-1}}$
13) 202 J
14) 58 J
15) $24.2\,\mathrm{m\,s^{-1}}$, $16.6\,\mathrm{m\,s^{-1}}$
    PE at D = PE at A
16) (a) no dissipative forces
    (b) (i) 0.16 J, $1.79\,\mathrm{m\,s^{-1}}$
       (ii) 0.12 J, $1.55\,\mathrm{m\,s^{-1}}$

**17)** 12.4 J

**18)** 118 J

**19)** 0.48 J. Energy stored in compressed spring

**20)** 5.57 m s$^{-1}$, 7.62 m s$^{-1}$ in same direction

## EXERCISE 5

**1)** (a) $114 \times 10^6 \, \text{N m}^{-2}$  (b) $6.22 \times 10^{-4}$

**2)** 39.2 mm

**3)** $80 \times 10^6 \, \text{N m}^{-2}$, $4.05 \times 10^{-4}$

**4)** (a) $0.01 \, \text{m}^2$  (b) 0.030 m

**5)** (a) $38.5 \, \text{MN m}^{-2}$  (b) $3.00 \times 10^{-4}$
(c) $128 \, \text{GN m}^{-2}$  (d) 628 N

**6)** $31.5 \, \text{GN m}^{-2}$

**7)** 41.2 mm, 0.87 mm

**8)** (a) $23 \, \text{MN m}^{-2}$  (b) 0.434 mm

**9)** 78 kN

**10)** 132 mm

**11)** (a) $25 \, \text{MN m}^{-2}$  (b) 500 kN

**12)** (a) $13 \, \text{MN m}^{-2}$  (b) $1.54 \times 10^{-6} \, \text{m}^2$

**13)** $200 \, \text{GN m}^{-2}$

**14)** (c) approx. $2.0 \times 10^6 \, \text{N m}^{-2}$
(d) $2.3 \times 10^{-3}$

**15)** $100 \, \text{GN m}^{-2}$

**16)** $200 \, \text{GN m}^{-2}$

**18)** Strong in compression

## EXERCISE 6

**1)** (a) 273 K  (b) 423 K  (c) 243 K
(d) 300 K

**2)** (a) $22\,°\text{C}$  (b) $-160\,°\text{C}$  (c) $-23\,°\text{C}$
(d) $127\,°\text{C}$

**3)** $216 \, \text{cm}^3$

**4)** $367\,°\text{C}$

**5)** $4 \, \text{N m}^{-2}$

**6)** $182 \, \text{cm}^3$

**7)** $0.374 \times 10^{-3} \, \text{kg}$

**8)** $8.3 \, \text{J mol}^{-1} \text{K}^{-1}$

**9)** $0.422 \, \text{m}^3$, 18.9 mols

**10)** $6.4 \times 10^{-5} \, \text{kg}$

## EXERCISE 7

**1)** 431 Hz

**2)** 285 m

**3)** $300 \, \text{m s}^{-1}$

**4)** $5.4 \times 10^{14} \, \text{Hz}$

**5)** $0.222 \, \text{g cm}^{-3}$

**6)** $85\,°$

## EXERCISE 8

**1)** $1.33 \times 10^{-13} \, \text{s}$

**2)** 400 m

**3)** $0.023\,°$

**4)** $14.5\,°$, $30.0\,°$, $48.6\,°$

**5)** $1.74\,°$

**6)** $22.95\,°$, 2, 5

**7)** 3rd

## EXERCISE 9

**1)** (a) $(3.1 \pm 0.1)$ cm  (b) 6 cm, 12 cm

**2)** Maxima and minima observed

**3)** 2.8 cm

**4)** $60\,°$

**5)** Visible, ultra-violet, infra-red

**7)** Matt black

**8)** To 'see' in the dark by detecting the warm areas of their prey

**9)** View through 'special' glass. Do not view directly

## EXERCISE 10

**3)** (a) (i) $37.5 \, \text{N m}^{-1}$  (ii) 0.56 s, 1.78 Hz
(b) 0.95 kg

**4)** 11.4 cm, 7.05 cm, $-7.05$ cm, $-11.4$ cm, 0 cm

**6)** (a) 0.020 s

**7)** (a) 0.20 s  (b) $\pm 29.6 \, \text{m s}^{-2}$, 0
(c) 0, $\pm 0.94 \, \text{m s}^{-1}$

**8)** (a) 0.89 s, 1.13 Hz  (b) $\pm 1.5 \, \text{m s}^{-2}$, 0
(c) 0, $\pm 0.21 \, \text{m s}^{-1}$

**9)** 19 kN

**10)** 1.40 s

**13)** (a) $944 \, \text{N m}^{-1}$

## EXERCISE 11

**2)** 0.25 T

**3)** $4 \times 10^{-5} \, \text{Wb}$

**4)** $1 \times 10^{-2}$ N

**5)** $3.0 \times 10^{-3}$ N

**6)** $3.0 \times 10^{-4}$ Wb

**7)** $2.4 \times 10^{-3}$ N

**8)** Parallel to field

**9)** $4.8 \times 10^{-6}$ N m

**10)** $B$, $A$, $n$ and $c$

**11)** (a) $B$ and $c$   (b) $B$, $c$ and $n$

**13)** $6.2 \times 10^{23}$ per m$^3$

**14)** $8.6 \times 10^{-14}$ N

## EXERCISE 12

**1)** $25$ mV

**2)** $5.0$ V

**3)** $0.3$ V

**4)** $4$ T

**5)** $3$ V

**6)** $4.7$ V

**7)** $8$ V

**8)** $20.8:1$

**9)** (a) $11.4$ kA   (b) $273:1$

**10)** (i) and (iii)

## EXERCISE 13

**1)** $3.7 \times 10^{-15}$ N

**2)** $7.4 \times 10^{16}$ m s$^{-2}$

**3)** $5.0 \times 10^5$ V m$^{-1}$

**4)** $4.0 \times 10^{-17}$ J

**5)** $1.39 \times 10^6$ m s$^{-1}$

**6)** $1.24 \times 10^4$ V

## EXERCISE 14

**1)** (a) $1.3 \times 10^{-15}$ N   (b) $1.6 \times 10^7$ m s$^{-1}$

**2)** B

**3)** $0.046$ m

**4)** $1.8 \times 10^{11}$ C kg$^{-1}$

**6)** $0.023$ m

**7)** $3.2 \times 10^{-19}$ C due to small number of drops observed

**8)** (a) $150$ kV m$^{-1}$   (b) $2.0 \times 10^5$ m s$^{-1}$

**10)** Each image of slit becomes blurred

**11)** (a) $0.29$ m,   (b) $0.31$ m

**12)** $5.4$ V

**13)** $17 \times 10^{-19}$ J

**14)** $6.2 \times 10^6$ m s$^{-1}$

# INDEX